"This is a must-read book about coming major disruptions in Capitalism that will rival the beginning of the Industrial Revolution in both impact and opportunity. Highly recommended  AGES OF READERS. In my opinion, two humanity faces today are climate change a Capitalism. This book provides a road-map urgently needed. A must read! – Bill Johnston, Former President of the NYSE, retired investor, CONCERNED grandfather

No futurist has been as focused on the climate crisis for as long or loudly as David Houle. The future David envisioned decades ago is now a reality that most of the world understands. The clear and present global danger demands that we shift from advocacy to activism. As David writes, "We can't flee… there's nowhere to go. Our fight is here and now." *Moving to a Finite Earth Economy* shares practical guidance for both individual and societal action. His innovative and brilliant solutions empower us to face down the existential threat of climate crisis through the prism of reinventing Capitalism. If you know someone, and we all do, who needs to move from advocacy to action on climate justice, share David's book and advice. – Jack Myers, Founder MyersBizNet/MediaVillage

"In *Moving to a Finite Earth Economy – Crew Manual*, Futurist David Houle takes a bold and courageous step to define the climate crisis not as one future trend of interest, but as the defining challenge of this moment in human history. We have 10 years to act, he argues. This trilogy offers a simple, clear, and prescriptive call to fight. It will challenge your complacency and any comfortable hope you may harbor that everything will work out fine through market forces or gradualism. Few thought leaders, and even fewer Futurists, are willing to lay it on the line like Houle does in this book. It may change everything." – Glen Hiemstra, futurist, author, speaker and founder of www.futurist.com.

"Houle has written a book that takes climate change literature to its next logical post-denial stage; that of prescribing what humanity needs to do to reduce fossil fuels by 2030. We must move to action and this book provides a road map for humanity. Highly recommended!" – John Englander, Oceanographer, Author of "High Tide on Main Street" http://www.johnenglander.net

"For those concerned about climate change and global warming, this book is a superb source of information and of something more – short, snappy sentences ready-made to be slogans, compelling slogans. It also shows you how global warming fits into the history of the planet, and how your everyday habits fit into the big picture." – Howard Bloom, philosopher, polymath and author. http://www.howardbloom.net/

"This is not a polemic from environmentalists asking you to please care about their pet issue. This is straight talk on what humanity needs to do to handle this existential challenge, undoubtedly the most important problem facing the world right now. Well-written, concise, backed up with sound research, *Moving to a Finite Earth Economy* tells it like it really is. It doesn't shy away from stating the hard truths of a climate-changed world, while at the same time recognizing that such a world, and our adaptation to it, contains significant opportunity for positive and fulfilling changes in our society." – Dr. Sean Munger, author "The Warmest Tide", historian, lawyer, public speaker

Moving to a Finite Earth Economy Crew Manual

the Complete Trilogy

David Houle

and

Bob Leonard

Dedication

To Victoria, with love, and gratitude for her patience and support.

To Christopher, may he continue to be the global citizen that we all need to become in the coming decades.

To Jordan may he thrive in the Finite Earth Economy.

To all those who are rising up with alarm at how humanity has not honored the planet and our future together on it.

David

To David for providing a platform for me to make my contribution to the struggle against this existential crisis, and for inviting me along on this adventure.

To Jamil Monroe who unwittingly changed the course of my life.

To Pamela Edwards and the other members of the PDX Pachamama Alliance for the work they are doing… and the community they are building.

To Adam, Alex, Anita, Beth, Ernie, Ken, Kerry, Patty, Paula, Sean and Vanessa who have made me feel at home in my adopted town of Portland, OR.

And to Buster, my little buddy.

Bob

Table of Contents

Acknowledgements

There are many people who have influenced us along the way in the writing of this book.

First and foremost, we acknowledge Tim Rumage, planetary ethicist and Head of Environmental Studies at Ringling College of Art +Design. David first met Tim when guest lecturing in Tim's classes at Ringling. That relationship developed into the co-authoring of "This Spaceship Earth" in 2015, and the co-founding of the global nonprofit https://thisspaceshipearth.org/ in 2016.

Bob came into the picture when he became the first Director of Operations of the nonprofit. Since then he has been promoted to Chief Content Officer. The purpose of This Spaceship Earth, the nonprofit, is to generate "crew consciousness" in as many people as possible. Neither David nor Bob would have the depth of insight, and the breadth of knowledge of the dynamics of global warming and climate change that infuses this book, without Tim's tutelage.

We want to thank all the people who have supported This Spaceship Earth over the last three years. Those who made financial contributions, those who participated in the many videos we produced, and those who attended our events and art projects. And those who shared photos and videos of them acting as crew members on Spaceship Earth.

We thank the environmental communities in both Sarasota and Portland for their support in our efforts to open people's eyes and minds concerning the unprecedented reality we all face.

David would like to thank Dr. Larry Thompson and the incredibly talented group of professors, staff and students at the Ringling College of Art + Design, who have embraced design and creativity just when it is most needed in the world. David would also like to thank the brilliant people who are members of The Sarasota Institute, a 21st century think tank... particularly his fellow co-founders Dr. Philip Kotler and Jason Apollo Voss.

We thank again all the great thinkers, scientists and leaders acknowledged in the Manifesto for a Finite Earth Economy for setting the stage in confronting humanity's greatest challenge.

We thank Mike Shatzkin, an advisor and friend, who has been instrumental in developing our go to market strategy first as an eBook trilogy, then this print version of the trilogy. He is the smartest guy we know concerning the future of publishing. Mike is also a leading advocate of both Carbon Fee and Dividend and for nuclear power as a necessary component to enable our transition off fossil fuels within the next 10 years. Mike introduced us to Phil Kahn, Co-Leader of the NYC chapter of Citizens' Climate Lobby and forward thinker concerning the oil and gas industries' potential leadership role in carbon sequestration (which we write about in detail in Chapter 11). Mike also introduced us to Herschel Specter, who spent much of his career in the higher echelons of the nuclear power industry. Herschel opened our minds to the role that nuclear energy can play in the Finite Earth Economy. He is quoted at length in Chapter 11.

In addition, our deep dive into nuclear power was assisted by William "Coty" Keller, PhD and retired Professor of National Security Affairs at the Naval War College. Coty supplied personal knowledge of the Navy's use of nuclear power.

We acknowledge the creative brilliance and ever supportive work of Dave Abrahamsen, a partner in the What2Design firm. He handled all of our design work on the entire trilogy, this book and the website that supports our efforts www.finiteeartheconomy.com Dave will continue to work with us as we educate as much of humanity as possible concerning the urgent need for a Finite Earth Economy by 2030.

We must also acknowledge Vanessa Benedetti. Vanessa lives one of the greenest family lives we know. She is the model of living with crew consciousness and has given us immeasurable insight into how to raise a family with a deep understanding of sustainability. She was a major influence for the "Personal" chapter of this book.

We also want to acknowledge Sean Munger, PhD for historical references that he shared to prove to us that humanity has faced existential threats before… and, in many instances, overcame them by banding together and refusing to give up. In these instances, societies often underwent massive changes in surprisingly short periods of time.

And we want to acknowledge the Pachamama Alliance for being an example of how to integrate indigenous spirituality into a crusade for environmental conservation… for being the change they want to see in the world: a collaborative community of activists.

Thank you all!

And Thank You! for buying this book. Help us activate with urgency the requirement to move humanity to a Finite Earth Economy by 2030!

Introduction

I urge all who read this to realize the magnitude of what we face and feel the urgency that we must face it with. Simply put, we have an opportunity to confront climate change, and move to an economic model that supports humanity by putting earth first... ahead of capital accumulation and economic growth. This will be an unprecedented undertaking, recasting the 225 years of Capitalistic Growth Economies that have run their course, to a Finite Earth Economy based upon removing greenhouse gas emissions as a necessary evil... a byproduct of our global economic systems.

Many economists, pundits, sociologists and even religious leaders are predicting the near-term extinction of Capitalism. We believe that the market-driven forces of capitalism can, after a challenging transition, supply the heart of a Finite Growth Economy with its new metrics. Growth economies require continual growth and we have reached the end of what the planet can support. We can't have infinite growth on a finite planet. A carbon-based, market-driven economy can work, but it will look quite different than the growth-based capitalism we are used to.

By moving from an economy based on capital to one based on reducing GHG emissions, from one based on growth and consumerism, to one based on zero growth and non-consumerism, we will be able to maintain quality of life while transitioning to a sustainable economic model. Market dynamics will need to change to the new reality we face. We cannot do this successfully unless we work together. Cooperation and collaboration must replace competition.

What you hold in your hands is the print compilation of an eBook trilogy that was published over a five-month period from April to August 2019. The reason to publish an eBook trilogy as a first step was to get the information out as soon as possible and to make each book a quick read.

In addition, the production and distribution of eBooks do not emit greenhouse gases. Many people are justifiably alarmed by the now tangible climate crisis and are doing anything they can to reduce their carbon footprint. We wanted to provide those people with the choice to buy and read Moving to a Finite Earth Economy with zero emissions. We believe that eBooks should always be significantly cheaper than print books, as they have minimal production and distribution costs. The eBooks sell for $2.99 each, or $8.97 for the full trilogy.

This print version is Print on Demand [POD] meaning that there are no advance prints runs, so no waste. Books are printed when they are ordered, as a single copy or in the hundreds. This minimizes the paper used and CO2 produced.

We want to lead by example. Slashing GHGs, is an overarching goal discussed in detail in the books. We must walk our talk.

In this book, we keep the structure of the trilogy (Book 1, Book 2 and Book 3) intact. The chapters are numbered the same as they were in the published eBooks.

Today in 2019, the silver lining (if one can call it that), is that climate change is HERE AND NOW. There is nothing in human genetics that triggers a response to something that might happen in the future. That's why there has been no global crusade taking swift and significant action to confront climate change. Today, a major swath of the world's population is experiencing the results of a warming atmosphere. This is good in that we are now in a 'fight or flight' situation. Our fight or flight response is genetically wired into us since pre-historic times and it is being triggered. We have no choice but to respond. So, humanity is ready to fight – to take meaningful action HERE AND NOW.

Most people in the world understand that climate change is real, and it is caused by humanity. As the general population has become more interested in the situation, mass media is giving it more coverage. Every day the number of people who think we need to "do something about climate change" increases. If you are one of those concerned humans, this book will show you what needs to be done and challenge you to take action.

This book will force you to realize how much you as an individual (plus your neighborhoods, cities, states, provinces and countries) must do to slow global warming – the catalyst of climate change. Are you ready to make the necessary personal and cultural changes, and become a postconsumer… even a non-consumer?

The Growth Economy has cast you in the role of consumer and our entire society is designed to support that. Are you ready to live with less? Are you ready to buy only what you need? Are you ready for massive change?

Are you ready to move to, work for and demand an economy that places Earth first? Are you ready to stop being a passive passenger on Spaceship Earth and take on the responsibility of being an active crew member?

If you accept that climate change is an existential threat, then this book will provide you with the ideas, metrics, milestones, technologies and policies at all levels to do your part in saving humanity. We are not "saving the planet", we are saving ourselves from ourselves. This is a self-help book for the human species… designed to prevent the collapse of civilization as we know it.

We need to do our absolute best as a species to move as much as possible to a Finite Earth Economy by 2030. Less alarming commitments with 2050 as the goal date are now too little too late… and they don't trigger the urgency of our fight or flight response.

In Chapter 7 we outline the various paths forward and conclude that anything later than the deadline of 2030 practically ensures the collapse of civilization by the end of this century.

Book 1 sets up the challenge at hand, the immediacy of NOW, the three things we need to do, and then examines the three types of economies: Growth, Circular and Finite Earth.

Book 2 conveys the new metrics, milestones and technologies needed to alter the trajectory of how humanity lives on Earth by 2030.

Book 3 outlines what must be done at every level, from personal to neighborhood to city to state to nation to global, including the actions required of corporations and governments.

NOTE: We have decided not to address the Green New Deal in this book. Our decision was based on the fact that the GND is a moving target. While we agree with many of its tenets, and there is some alignment with FEE, nobody knows what the GND will end up being, so we have chosen not to incorporate it.

The last important point to make is that this trilogy is written by a futurist.

Most climate change books are written by climate scientists, activists or ecologists. A futurist brings a broader view of all that relates to climate change, to all the trends that are not about climate change, but which are extremely relevant in addressing it, or will be impacted by it.

As a futurist I hold as my highest responsibility the delivery of accurate forecasts. That is my value to the world. My accuracy has been very high. I have been wrong mostly on timing and scale. If you want to verify that statement, please go to www.davidhoule.com to see my past forecasts.

In "Moving to a Finite Earth Economy – Crew Manual" we make it totally clear that the time is NOW, as what we face is the single largest endeavor humanity has ever undertaken.

Thank you for buying this book. We hope it opens your mind, changes your thinking and prompts you to act differently, and to do so with urgency. Please enjoy this book. Learn. Share. Tell others. Give us a review. Most importantly, be ready to act with greater knowledge, concern and commitment.

David Houle and Bob Leonard

Moving to a Finite Earth Economy

Crew Manual

Book One - The Three Economies

Chapter One

The Biggest Challenge
We Have Ever Faced

"In this early part of the 21st Century, we are in a planetary reality for which we have no precedence. The last time the atmosphere had as much CO2 as today was at least 800,000 years ago. Modern Humanity has been around for 200,000 years so there is no road map, plan or strategy we can pull up from history." — Houle and Rumage, "This Spaceship Earth"

Facing climate change and slowing global warming is the biggest challenge humanity has ever faced. It is without precedence.

Global warming and the resultant climate change now occurring has never happened since humanity has been on earth. There is nothing in our experience or our DNA to prepare us. The natural cycle of CO_2 increases and decreases, going back hundreds of thousands of years, has been broken. This means that everyone, scientists included, are making forecasts and looking ahead with less certainty than we all would like.

What is also clear is that climate change is no longer a future threat, it is a present reality. What humanity has avoided facing over the last 50 years is here... now. Since we have not been proactive in the past few decades, we now urgently need to make major changes. We must be both immediately reactive and long-term proactive if we desire human civilization to avoid collapse.

In This Spaceship Earth (co-authored by David Houle and Tim Rumage, and published in 2015), the case was made that humanity must, by 2030, dramatically change the trajectory of how we live on Earth. Or we face the collapse of civilization by 2100. Even four years ago, that view was thought of as alarmist. Now it is our reality. Confirming David and Tim's forecast, the IPCC report of October 2018 states the urgent need to make massive changes in almost every facet of modern life by 2030.

R. Buckminster Fuller in his two books of the late 1960s and early 1970s: "Operating Manual for Spaceship Earth" and "Utopia or Oblivion: the Prospects for Humanity" set forth the clear view that if humanity did not create whole systems processes for operating Spaceship Earth, then decades later (now), we would be unprepared for the inevitable fork in the road: utopia or oblivion. It is clear that we are rapidly advancing down the path to oblivion. We need to halt our progress toward oblivion, so that we can have a chance in the future for utopia.

In 2018 the perception of climate change shifted. Alarm bells were going off all around the world. Unprecedented fires, floods, droughts, storms and air pollution finally provoked many more people, politicians, NGOs and corporations to call for action.

The Intergovernmental Panel on Climate Change (IPCC) issued a report in October 2018 stating the dire need for action by 2030 since the earth will warm by 1.5 degrees Celsius by then[1]. It went on to state that if significant efforts to reduce greenhouse gas emissions are not taken by 2030, climate catastrophes would grow in number and severity. The message is that, to prevent disastrous weather events, droughts, floods and famine, humanity must not allow warming beyond 1.5 degrees C. Earth has already warmed past one degree C since pre-industrial times 1750-1800.

The Purpose of this Book

The purpose of this book – and the two that follow – is to outline required actions to slow global warming and avert catastrophe. People want to do something. These books provide the action items and policy suggestions at all levels of humanity, from personal to global.

The magnitude of the effort required may shock you. So be it. It's time to quit talking about it and go to work. If we don't, we won't have a second chance to walk back from Fuller's oblivion path and save civilization. We may be able to take the path toward utopia sometime in the future, but first we must save ourselves from ourselves.

The operative word in confronting climate change is URGENCY.

The fact that climate change is already wreaking havoc, can be viewed as a positive. We'll discuss why in Chapter 2.

Takeaways

- Climate change is the greatest challenge humanity has ever faced, and we are in uncharted territory.

- We must launch massive changes in how humanity lives on Planet Earth well before the 2030 deadline.

- If humanity does not make these massive changes, there will no longer be civilization as we know it by the end of this century.

- We must declare a state of urgency.

Chapter Two

Immediate and Present Danger

"Adults keep saying, 'We owe it to the young people to give them hope.' I don't want your hope. I don't want you to be hopeful. I want you to panic. I want you to feel the fear I feel every day. And then I want you to act. I want you to act as you would in a crisis. I want you to act as if our house is on fire. Because it is." — Greta Thunberg, 16-year-old climate activist

Climate change is no longer a future threat. It is now a present reality.

We have already passed the fork in the road of R. Buckminster Fuller's 'Utopia or Oblivion' and mindlessly gone down the wrong path. But we didn't know it. Now we do.

At this juncture, in 2019, we face two futures. We dramatically reduce our use of fossil fuels and resource consumption by 2030… or we do not. If we do, we will still spend decades on a warming planet and live in a less than optimal reality; one less pleasant than what we are used to. If we don't, civilization as we know it will no longer exist by 2100, maybe earlier.

As mentioned in This Spaceship Earth, published in 2015, the book's authors, David Houle and Tim Rumage, have been clear that it is 2030, not 2040 or 2050 that is the critical benchmark year. Just four years ago, that view was thought of as alarmist. Now it is our present reality.

The lack of true understanding of the magnitude and immediacy of climate change shows up when any person, group or organization references 2050 as a deadline for achieving any level of reduction in greenhouse gas emissions. As the quote that opens this chapter suggests, our home is on fire. Though this heating of the planet is perceptually moving at a slower pace than a house on fire, the fuel (greenhouse gas emissions) continues to accumulate in the atmosphere and in the oceans. By 2030, we'll reach a tipping point resulting in a runaway hot house state.

If all greenhouse gas emissions stopped today, we would still have two to three decades of warming due to the resident CO2 already in the atmosphere. Pre-Industrial Revolution, there were 730 gigatons of CO2 in the atmosphere. Today we are at 1300[1]. So, in addition to moving to a Finite Earth Economy, we must develop and use new (and yet to be invented) technologies to extract greenhouse gases from the lower and upper atmospheres. A safe concentration would be about 800 gigatons as the goal, with 500 gigatons removed as fast as we can.

A primary reason for our lack of urgency in dealing with this existential crisis is that humans are not wired to aggressively respond to a potential future threat. We respond with urgency when the threat is immediate and apparent.

Climate change in 2019 is an immediate and apparent threat. We have no time left to waste. Those alive now must alter course and move to a Finite Earth Economy for our own quality of life, and so future generations have a habitable planet to live on. Our inertia in facing climate change over the last 70 years has backed us into this corner.

Fight or Flight

Our current urgent state of affairs, oddly, has a silver lining.

Humans have never been good at anticipating future problems... why look for trouble? In 2019 the climate threat is now. Our fight or flight response is being activated.

From the time we were hunters and gatherers, we have had an embedded genetic program running called "fight or flight". When confronted with a real immediate danger – a bear, an avalanche, a deadly snake – this program triggered one of two responses: we confronted and fought, or we ran. We do not have genetic programming that triggers any such reaction to a potential threat that might happen sometime in the future.

Since Alexander Graham Bell first raised his concern of global warming back in 1917, it has been a potential threat that might happen at some time in the future. Collectively we became aware of global warming, climate change and environmental issues at the first Earth Day in 1970 – 50 years ago.

In the years since, there have been many concerned climate scientists, such as James Hansen, Katharine Hayhoe, Bill McGibben, Michael Mann and many others who have used hard, well-researched, peer reviewed science to publish warnings that humanity was heading toward a climate catastrophe. In 1988 Hansen, perhaps the foremost scientist sounding the alarm, published his 'Global Climate Changes as Forecast by Goddard Institute for Space Studies'[2]. In it he warned of catastrophic global warming caused by greenhouse gases. That was 30 years ago, and things have only gotten worse.

Almost every scientific paper published forecasts a climate crisis. All of them have had to amend their prior reports as being... TOO CONSERVATIVE! In every case the reason is that the pending calamity is not moving in a linear fashion, but on more of a geometric acceleration. Realities forecast for 2025 were happening in 2000. The initial projections for 2050 are happening now.

Why? No human who has ever lived has experienced the planetary reality unfolding now. It has never happened to modern humanity, so we have no frame of reference, no context, no history to point to. Even the most brilliant climate scientists can only make estimates based upon experienced reality and extrapolate from there.

"I've never been at a climate conference where people say, 'that happened slower than I thought it would'." – Christina Hulbe, Climatologist

The second reason is that our brains aren't wired to react to a future event that has no empirical grounding in the present. Now that we see the clear and present danger, now that we are experiencing massive forest fires, monster hurricanes, record breaking droughts and more, we see the need to act. Our fight or flight responses are kicking in. We can't flee... there's nowhere to go. We must fight. Our fight is here and now.

All of humanity, as many of the 7.7 billion of us alive as possible, must react and react urgently with everything we have to this immediate and present danger... this existential threat.

This book, and the two that will soon follow, will explain what needs to be done. What humanity must do from bottom up and top down. What countries, corporations, provinces and states, cities and towns, neighborhoods and individuals must do. People who have wondered what they might do, or felt frustrated because the problem is so gigantic, will be given a roadmap, a blueprint for taking action in meaningful, doable ways. It won't be easy, but we have no choice. It will certainly challenge all of us... it may overwhelm some. And we need to get it done by 2030.

Three Important Concepts

In This Spaceship Earth David and Tim stated three things that must be understood by any of us who want to act:

Forgiveness

When we find ourselves in a bad situation of our own creation, we often look to place blame. We blame ourselves or others. Let's not waste time and energy on that. Let's forgive ourselves and forgive others. Relationships that have been damaged can only be renewed if one starts with forgiveness. We have a massively damaged relationship with the only place we live – Spaceship Earth.

We must forgive ourselves for not having acted sooner... for mindlessly burning fossil fuels. For not tying our actions to their consequences. For buying more than we need in our advertising-driven, throwaway consumer economy. We were born into and grew up in this culture, we did not create it. So, let's proceed with a collective forgiveness for all that we have and have not done.

"Save the Planet"

Anyone who uses this phrase is showing their ignorance about Climate Change. We see ads that tell us to buy a product to 'help save the planet'. And charities ask for donations to 'help us save the planet'.

That is a huge misconception! How anthropomorphic for us to assume that.

Planet Earth is some 4.5 billion years old. It will be around for billions of years more. To put things in perspective, humans have existed for 200,000 years. That's 0.00004 percent of the total life of the planet.

"We need to save ourselves from ourselves." – Tim Rumage, planetary ethicist

What Tim means is that climate change is caused by humans – by our consumer society and an economic system that requires constant growth. We extract natural resources, manufacture goods, use them and dispose of them. The waste is enormous and many of the byproducts are toxic. The buildup of greenhouse gases generated by man is causing the warming of our atmosphere. We are the cause of the problem and we must be the solution. If we don't change how we live, humanity will go extinct (potentially as soon as the end of this century). Many other species will also go extinct. And within five million years or so, Earth will have recovered, and new species will have evolved.

What needs to be "saved" is us.

The Cause of Climate Change

How we live on Earth has triggered what we are now experiencing. This is largely due to our ignorance of the consequences of our actions. We don't know that a quarter-pound hamburger (from ranch to plate) emits six pounds of CO2. That the T-shirt we're wearing required 80 gallons of water to manufacture. We think and live in silos, seldom reflecting on what may be affected in adjoining silos, or in the entire system.

The reality of living on Earth is that everything is interconnected. Our siloed thinking has created a disconnected view of cause and effect, where the actions we take are separated from the consequences they cause.

"Belief in separateness is the root of all suffering." – Buddhist saying

We must examine our actions, with an understanding of interconnectedness.

Takeaways

- Climate Change is happening NOW.

- Our collective delayed response to global warming means that we are no longer discussing some vague future event. Our fight or flight response is being triggered. It's happening here and now!

- We must start by standing in forgiveness. We must forgive ourselves as we were all born into a Growth Economy/fossil fuel reality. We didn't have any real choice. We do now and we must make the right one.

- The planet doesn't need saving. It is simply adapting to the warming that humans have created. We need to save ourselves from ourselves.

Chapter Three

The Three Things Humanity Must Do

*"The approach to resolving Climate Change is to focus on the reduction of CO2 emissions. It is an understandable yet insufficient approach. There is a lesson from childhood that applies to this topic, but it is a tough lesson to learn. The lesson is: **Being less bad is not the same as being good.**"*

– Houle and Rumage, This Spaceship Earth

The question we are most often asked is, "What can I do?"

Over the last few years, climate change has become a reality. The evidence is everywhere. The majority of people are finally saying, "We have to take action."

Our collective fight or flight response has been triggered. We can't escape it, so how do we fight it?

The magnitude of our situation is larger than any other challenge that humanity has ever faced. And the scope of potential actions is almost beyond comprehension. We must make changes in practically every area of human endeavor. How we live as individuals, and in our neighborhoods, towns and cities, states and provinces, nations and as a species must all evolve beyond the status quo. Our current lifestyle has caused the problem.

This book will deal with specific actions and policy suggestions for all levels of human society and economics. We will examine three types of economies and determine which has the best chance of driving the changes we need to make.

Three Big Things Humanity Must Do

On a macro-level there are three categories of collective action that absolutely must happen if we are to halt the warming that results in climate change.

1. Eliminate the burning of fossil fuels as quickly as possible, and significantly lower the rate at which we are consuming earth's resources.

2. Draw down the greenhouse gases already in the atmosphere.

3. Create "crew consciousness' in as many of today's 7.7 billion passengers on Spaceship Earth as possible.

Let's consider each of these three in detail:

Reduce our use of fossil fuels as fast as possible. At the same time, reduce all levels of consumption.

This is what most people think of when they consider our effort to conquer climate change. Almost everything that is promoted or discussed in the media fits into this category. And it is a crucial component… absolutely essential to the success of our effort. And it will require the redesign of the global economy.

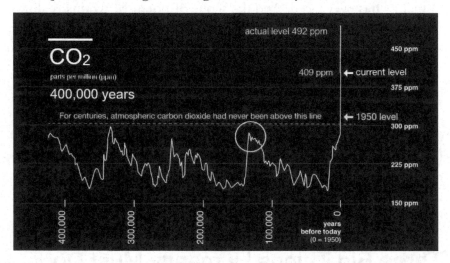

Chart 3.1 - Slide from Paul Hawkens' Drawdown presentation (see Addendum for other charts)

This chart documents that, since 1950, the Parts Per Million (PPM) of CO2 in the atmosphere has increased by over 60% to its highest level in at least 400,000 years!

The 492 ppm depicted includes Methane (CH4). While CO2 can remain in the atmosphere for a century or more, CH4 dissipates in about a decade. But CH4 is 30 times as effective in trapping infrared radiation (the Greenhouse Effect) as CO2[1].

The following chart shows the amount of CO2 humanity has emitted into the atmosphere annually over the past 50 years. Along with the PPM measurement, this is the other metric (gigatons emitted into the atmosphere), that we must pay close attention to, because it is this number that must be reduced by 50% by 2030, if we are to avoid catastrophe.

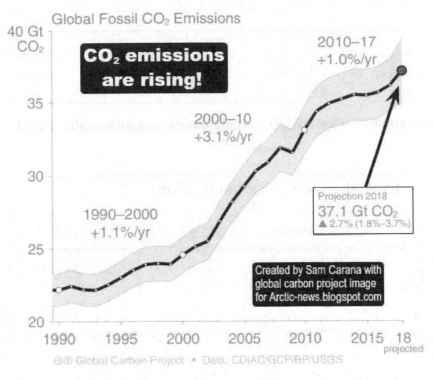

Chart 3.2

We must reduce the PPM number by 20%, and the gigaton emissions level by 50% by 2030.

Because CO2 is transparent... we can't see it, it's difficult to visualize how much is being released. To get a better picture of how much CO2 humanity is pumping into the atmosphere each year, imagine an African elephant that weighs four tons.

Now imagine releasing 300 of these elephants into the atmosphere (imagine they can float) **every second of every day for an entire year.** That's how much CO2 we put into the atmosphere (37 gigatons) in 2018. Imagine looking up at the sky every day and seeing thousands more elephants floating up in the atmosphere.

Note: The COP meetings, the Paris Climate Accord and other initiatives are planning to cut emissions by the year 2050. That milestone is no longer supported by science. On humanity's current trajectory, the biosphere will have warmed far beyond 1.5C by 2050, mostly likely 2-3C, with 2100 being a 5-7C increase. The end of civilization as we know it will start to occur at 3C.

So, the deadline for everything in this book is 2030. To the very best of our collective and collaborative capabilities, that is the goal if we want our descendants to have anything better than a nightmare existence.

2030! 2030! 2030! 2030! 2030! 2030! 2030! 2030!

Reduce all levels of consumption.

We learned about the three R's: Reduce, Reuse, Recycle on the first Earth Day in 1970. Unfortunately, we did nowhere enough reducing, reusing and recycling to make a difference. We will discuss the Circular Economy in depth in Chapter 6.

"While the U.S. produces around **236 million** tons of municipal solid waste every year, the numbers for industrial waste are far less clear. Some estimates go as high as **7.6 billion** tons of industrial waste produced every year." – RecoverUSA, Waste Stream Statistics[2]

Around the world, waste generation rates are rising. In 2016, the worlds' cities generated 2.01 billion tonnes of solid waste, amounting to a footprint of 0.74 kilograms per person per day. With rapid population growth and urbanization, annual waste generation is expected to increase by 70% from 2016 levels to 3.40 billion tonnes in 2050. Source: World Bank[3]

The problem is that humanity is consuming more than Earth can regenerate. The Global Footprint Network has come up with a unique way to measure our consumption using the concept of Earth Overshoot Day.

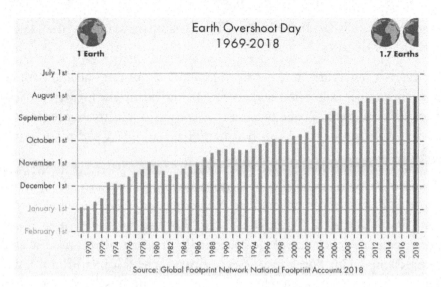

Earth Overshoot Day
1969-2018

1 Earth   1.7 Earths

Source: Global Footprint Network National Footprint Accounts 2018

Chart 3.3

This chart clearly shows that starting in 1970 (the year of the first Earth Day) humanity has increasingly consumed more than the planet can sustain. As of the writing of this book, humanity consumes at the level of 1.7 Earths per year. This is an enormous problem. When you think about the aggregate overages in the last 48 years, we have dug ourselves a huge hole.

In This Spaceship Earth, we wrote that the only valid definition of the word "sustainability" was at the level of the planet. Therefore, we have consistently used the concept of Earth as a spaceship. A spaceship is an entity traveling through space that has a finite amount of resources onboard and must live within those confines. What would happen to the crew aboard a spaceship who were consuming at a 70% higher rate than could be regenerated? The mission would have to be aborted.

As a spaceship must operate within its finite on board supplies, so must Earth be operated within the planet's finite resources. At least since 1970, we have steadily been depleting those resources. Planetary sustainability means operating Spaceship Earth at a 1.0 or less Earths per year. This means that Earth Overshoot day should be no sooner than 12 midnight on December 31. Since, for half a century, humanity has extracted more of the Earth's finite resources than the planet could regenerate, we will need to compensate so that Earth Overshoot Day is sometime in February or March of the following year. And then keep it there for decades.

In the past few years there has been an explosion of online and offline content concerning living with less, getting rid of stuff, buying used not new, simplifying our lives and lowering our carbon footprints. Living with less is a concept that many have adopted, and many have found it freeing.

The developed economies are now moving from an ownership paradigm to one of rental/access. We don't buy DVDs anymore, we rent from Netflix. It's the same with CDs and Spotify. The average American uses their car 4% of the time. The rest of the time it sits depreciating. This doesn't make economic sense. More and more, we use public transportation where we can, we rent cars by the hour or day, and we Uber, Lyft and ride share.

The sharing economy is ascendant and just in time for the shift this book suggests. Moving from an ownership to an experience economy is a very supportive dynamic for a Finite Earth Economy

Fossil Fuel Use and Consumption

In summary, we must reduce our fossil fuel use globally by at least 70% by 2030, and we must reduce our consumption of goods and products globally by 50% by 2030. It won't be easy, but it can be done.

Ultimately all reductions only work at the global level. Certainly at least the 20 largest polluting economies of the world must lead... and do so aggressively. Climate change does not recognize national boundaries.

> *"Everyone is downstream and downwind from someone."*
> Houle and Rumage "This Spaceship Earth"

Reduction of fossil fuel use is a primary requirement in meeting the challenge of climate change. There are two additional undertakings that are essential if global warming is to be slowed.

Drawing Down of Greenhouse Gases from the Atmosphere

When CO2 enters the atmosphere, it stays there for a long time. It can remain in the atmosphere for centuries. That's why it accumulates, and the PPMs keep rising.

The retained and growing atmospheric CO2 triggering warming is called Resident CO2. This is a critical point to understand. Ending emissions is a primary focus, but without a drawdown of what we have already emitted into the atmosphere for the past 75 years, the planet will continue to warm for decades.

Prior to the start of the Industrial Revolution, the estimated amount of gigatons (one billion metric tons) in the atmosphere was 730. During the 1960s and 1970s, the number passed 800 gigatons. We are now close to 1300 gigatons, an increase of 500 gigatons in 38 years. This is simultaneous with the warming of the surface of planet Earth.[4]

This first chart shows the increase in gigatons of CO2 in the atmosphere.

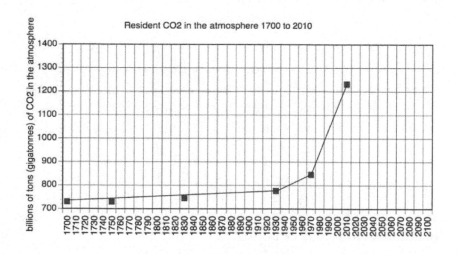

Chart 3.4 - This Spaceship Earth, *Houle and Rumage*

This chart shows the increase in annual surface temperature of Earth since 1880.

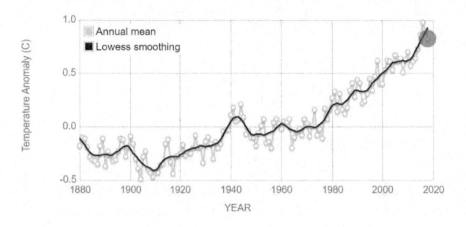

Chart 3.5 - NASA's Goddard Institute for Space Studies (GISS)

Note the congruency of the upward curve in both charts starting around 1970.

Apart from 1998, 18 of the 19 warmest years all occurred since 2001. 2016 ranks as the warmest year on record.

We are a 730 gigaton species living in a 1300 gigaton world. It is this 500+ gigatons of CO_2 in the atmosphere that has triggered the global warming of the last 40 years. It is this additional 500 gigatons that will trigger continued warming even if all emissions are halted. In order to slow global warming, we must extract as much of those 500 gigatons as possible, as soon as possible.

Currently this is prohibitively expensive, but as with all technologies, the price will come down and the functionality will increase through innovations and economies of scale. We will discuss this in detail in the Technology chapter of Book 2.

Creating Crew Consciousness

This is the third action we must take to slow global warming and successfully tackle climate change. Most of humanity thinks and acts as passive passengers on Spaceship Earth. They have no awareness of, or regard for, the finite supplies stored on the planet. Throughout history, until very recently, the planet easily provided as much as we needed. We now realize we have plundered its capacity to support life.

"We need to move from being unaware passengers on Cruise Ship Earth to being aware crew members of Spaceship Earth." – planetary ethicist Tim Rumage

The two greatest thinkers concerning the concept of crew consciousness are R. Buckminster Fuller and Marshall McLuhan:

"There are no passengers on spaceship earth. We are all crew." – Marshall McLuhan

"Spaceship Earth was so extraordinarily well invented and designed that, to our knowledge, humans have been on board it for two million years not even knowing that they were on board a ship." – R. Buckminster Fuller, "Operating Manual for Spaceship Earth" *more quotes from this brilliant thinker in the Addendum

In the Age of Climate Change, we must step up to crew our planetary spaceship… the only home we have. As stated at the opening of this book, what is happening now in the atmosphere and biosphere has never happened before during man's time on the planet.

Tools defined the Agricultural Age

Machines defined the Industrial Age

Technology defined the Information Age

Consciousness will define the Shift Age.

David Houle, The Shift Age, 2007

To face a completely new planetary situation, we need to start with a new consciousness. A new consciousness leads to new thinking, which leads to new actions. If we accept full responsibility for crewing Spaceship Earth, we will change what we do and how we do it, both separately and together. Scaling up as crew means that individual actions will matter, will produce results as our crew numbers grow from millions to billions. (This is why we set up the nonprofit This Spaceship Earth to help create crew consciousness on Spaceship Earth.)

Let's illustrate what crew consciousness means. When you receive your monthly electric bill, you have a certain price expectation. The average American's electric bill is a bit more than $100 which pays for an average of 1000 kwh/month.[5]

If we asked you how much your monthly electric bill is, you'd be able to tell us. If we then asked you how much electricity you use per month, you'd likely have no idea. Crew consciousness changes that. It makes the primary metric not the cost, but the amount used. When you know how much you're using, it becomes easy to track it and reduce it… to make sure you aren't using more than you need.

If just 25% of homes in America and other developed countries acted as crew with their electric bills and consciously lowered their electricity use (a majority of which is generated by burning fossil fuels) there would be a significant decrease in greenhouse emissions.

Crew consciousness is first, understanding that you are a crew member, not a passive passenger. Second, becoming aware of the level of your resource consumption. Third, determining if you need to reduce your resource consumption. If you do need to reduce your consumption (and most of us in the developed world do), you would then take the necessary steps to accomplish that reduction.

Active and conscious stewardship of the resources we each use is crew consciousness. Crew consciousness is a crucial component of moving to a Finite Earth Economy. In Book 3 we will lay out what crew consciousness might look like at every level of society.

We end this chapter with the three numerical targets that were in the "Moving to a Finite Earth Economy – Manifesto"

By 2030, we need to:

- move to 70% renewable and alternative energies globally,

- reduce atmospheric CO2 from 420 parts per million to 315 ppm,

- cut the current 37 gigatons of annual CO2 emissions by at least 50%.

Takeaways

There are three things humanity must do to begin to slow global warming:

1. Decrease the burning of fossil fuels as quickly as possible, and significantly lower the rate at which we are consuming earth's resources.

2. Draw down the greenhouse gases already in the atmosphere.

3. Create "crew consciousness' in as many of today's 7.7 billion passengers on Spaceship Earth as possible.

While #1 is mission critical, #2 and #3 are also essential.

Chapter Four

The Three Economies

"Anyone who believes exponential growth can go on forever in a finite world is either a madman or an economist." – Kenneth Boulding (economist)

"Environment and economy are two sides of the same coin. If we cannot sustain the environment, we cannot sustain ourselves." – Wangari Maathai

"Entering the 21st century, the vast majority of humanity has been living in a world with a footprint greater than the planet's capacity. Clearly, humanity cannot continue to live beyond the Earth's means... we must alter the trajectory, the thinking, and the practices that have placed us in this mission critical state." – Houle and Rumage, This Spaceship Earth

Here we are, 19 years into the 21st century. We have postponed preparing for global warming and climate change for 50 years. We are now confronted with the reality that both are here, now. We put ourselves into this current position of "fight or flight".

I have long thought that the 21st century would be the Earth Century. It has been a major concept in my presentations for the past decade. This is the century when we collectively prioritize our relationship with planet Earth above all else.

"The twenty-first century will be the Earth Century. It will be during this century that humanity faces the reality of whether it wants to destroy itself, and much of what exists on this magnificent planet, or not. Assuming we make essential course corrections, future historians will write about the Earth Century as a turning point in human history… So stop getting excited about Earth Day. Retire that thinking and refocus on the next ninety years of the Earth Century." – David Houle, "Entering the Shift Age" 2012

As a futurist, I have stated that we left the Information Age and entered the Shift Age. Ranging roughly from 2005 to 2030, this is when much of human reality shifts to the new, future reality. It's a transition from what was to what will be.

"The Shift Age is and will be the age when humanity alters its sense in relation to our planet. This goes way beyond what we think of as the environmentalism of today. It is a much more long-term stewardship point of view. All of our previous history has been about economic growth and use of the planet's resources for our immediate needs. The Earth seemed to be unlimited in its space, resources and ability to absorb whatever humanity did. No longer." – David Houle "Entering the Shift Age" 2012

How will humanity be living at the end of the Earth Century? Will civilization be at the same level as we have now? In a basic and existential way, the answer is truly up to us. Our future is ours to design, create and live.

"We are called to be architects of the future, not its victims." – R. Buckminster Fuller

A huge part of shaping our collective future will be how we change the global economy and all our respective nation state economies. Driven by the Carbon Combustion Complex, the growth economies of today must cede to a new economic model.

In the Earth Century, humanity will have to move away from our growth economy paradigm. There are alternatives to the growth economy. We focus on two that are designed to slow our destruction of the biosphere. They are the Circular Economy and the Finite Earth Economy.

The chart below frames all three:

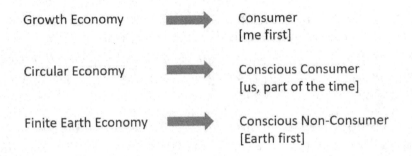

Growth Economy → Consumer [me first]

Circular Economy → Conscious Consumer [us, part of the time]

Finite Earth Economy → Conscious Non-Consumer [Earth first]

Let's look at each of the three economies.

Growth Economy

The American Dream has been exported around the world. It is the dream for more: more cars, more factories, more buildings, more freeways, more money, more houses, more appliances, more everything. The dream has become a nightmare, wreaking havoc across the globe.

Every reader of this book has lived his or her life in a Growth Economy. For most of history, humans have lived in growth economies. Think about how the economic health of any developed country is measured. All the metrics concentrate on growth, expansion and output. Gross Domestic Product (GDP) is an output. We measure the success of a nation's economy by how much growth has occurred in its GDP.

In all of these growth economies, there is little or no accounting for the use of natural resources. Countries behave as though Mother Nature provides its bounty for free. Our current form of Capitalism does not take into account the value of natural capital.

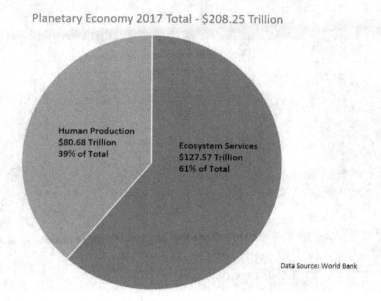

Chart 4.1

It is time to redefine global GDP by taking a more planetary approach, one that values Nature's resources. More on that ahead when discussing the Finite Earth Economy.

Currently, natural capital is not accounted for in the financial reporting of Growth Economy corporations. And the majority of GDP (measured in dollars) is purchased by individual consumers. In the United States, the world's largest economy, consumer spending accounts for 70% of GDP. Our consumer society urges us to shop, buy, buy, consume, buy more. Keep up with the Joneses. Your self-worth is determined by how much stuff you accumulate. When a product begins to wear… throw it out. Buy another one.

This is now a major problem because we are consuming more than Earth can regenerate. Americans' annual consumption is at the rate of five Earths. In other words, if all the world consumed at the American level, five Earths would be needed to support our current lifestyle.

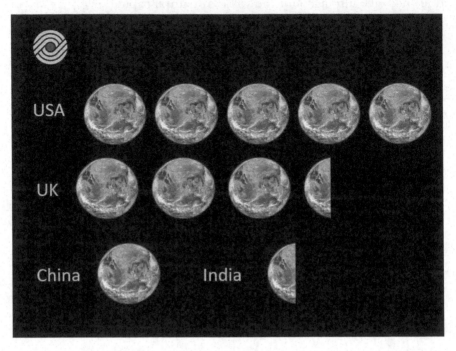

Chart 4.2 - Per capita consumer levels Source: Global Footprint Network

We have all been raised in this economic environment… where our personal consumption is disconnected from the depletion of planetary resources. We have been born and raised in a world saturated by advertising. We receive hundreds of messages every day, via every conceivable format, designed to persuade us to buy ever more things. As stated in Chapter 2, we should forgive ourselves for our planetary transgressions. We didn't know that how we were living, collectively and individually, was stressing the planet. If there were only one billion people on Earth, it could support our consumer lifestyle… but there are close to eight billion people with more being born every minute.

So, we start with forgiving ourselves for what we have unknowingly done. Now that we know – know that we are at cause for global warming and climate change – what are we going to do?

We need to move away from our Growth Economy model as fast and as far as possible.

The Circular Economy

Environmentalists realized decades ago that our Growth Economy was the engine driving pollution, biosphere degradation and global warming. With the first Earth Day in 1970, an ever-increasing population of environmentalists decided to create an alternative economy. They named it the Circular Economy. The three Rs – Reduce, Reuse, Recycle – became a part of our lexicon. It espouses an economy that was to replace the produce, use and discard paradigm with a closed loop system that produced, used and then reused or reprocessed. The result would be lower consumption.

There are many smart climate scientists and environmentalists who still think we should embrace the Circular Economy. It has cachet in some circles as a viable alternative to the Growth Economy model. The issue is that, as of this writing, the size of the circular economy is roughly 9% of the total worldwide GDP.[2] It is now 49 years after the first Earth Day in 1970, when the underlying concepts of the Circular Economy first became widely known and understood. If we had collectively stepped up and committed wholeheartedly to the development and expansion of the Circular Economy, perhaps half of the global economy would today be circular. Our climate change plight would be much less dire.

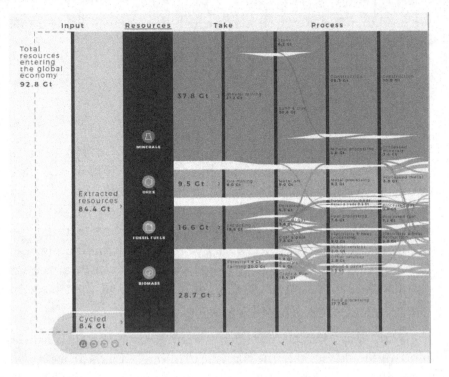

Chart 4.3 - Source: The Circularity Gap Report, Circle Economy, January 2019

The Circular Economy would have been a great alternative to the Growth Economy, if a large percentage of the population had embraced it, but they didn't. Now, almost 50 years from that first Earth Day, only 9% of the global economy can be categorized as circular. This is another example of our choosing not to take action because climate change was perceived as a future, not a present threat.

We have squandered the past several decades and humanity has not collectively adopted a Circular Economy. And now it's too late. It took 50 years to reach 9% of worldwide GDP.

We have 11 years.

Finite Earth Economy

As the chart at the top of this chapter notes, we must move to an economy that is about conscious non-consuming. We must create an economy that puts the Earth first. It is about us and how we manage the finite resources of the planet. We must save the only home we have. We must move to a Finite Earth Economy.

In a Finite Earth Economy, Earth comes first. All economic, material and energy issues attend to the needs of the planet first… then humanity's needs.

Earth First.

The premise of this book, and the reality of the urgency of our current situation, is that we must move to a Finite Earth Economy as quickly as possible.

Takeaways

- The 21st Century is the Earth Century. This is the century when humanity
  MUST deploy a sustainable stewardship of planet Earth.

- The Growth Economy model, which has served humanity for all of history, must now be retired as unlimited growth cannot be sustained on a finite resource planet.

- The Circular Economy never gained enough traction to counterbalance the damages done by growth economies – particularly since the global population has tripled over the last 70 years.

- Moving to a Finite Earth Economy is essential for slowing global warming, confronting climate change, and securing the long-term economic viability of human societies.

Chapter Five

The Growth Economy
(the Problem)

"Only in economics is endless expansion seen as a virtue. In biology, it is called cancer." – David Pilling

"In a consumption based economic model, creating waste is automatic." – Houle and Rumage, This Spaceship Earth

We have all grown up in a Growth Economy. It's the only way of life we've known. It is the economic reality that has shaped all our lives. Our reality equals living and participating in a Growth Economy. So, it will require a massive effort to transform our current consciousness (beliefs about many things) and how we behave in the world to create a different economic order.

Reality for most people is what they have experienced and learned in their lifetime, plus beliefs and behaviors installed in them by their parents, teachers, etc. To be open to see the future requires that the present perception of reality be, at least temporarily, suspended.

The only constant in the universe is change. Reality as we know it is always changing. In the Shift Age, the speed of change has accelerated to the point that it is no longer an accelerating line of change. It is environmental: we live in an environment of constant change.

We must suspend our perception of reality when dealing with Growth Economies. We must all step out of "that's just the way it is" to see how our collective agreement regarding economic reality is, in fact, the primary cause of global warming and climate change.

We must decide that the Growth Economy is not the only way forward, though it is the only way we know. If we want there to be some remnant of "civilization" in the year 2100, we must transition away from the Growth Economy model.

Until the second half of the 20th century, Growth Economies expanded and accelerated, without any perceived negative planetary consequences. There was no discernible realization in capitalist countries that growth was anything but good. There were no visible major negative effects. There were reasons for this. First, the number of people living was a fraction of what it is today. Second, productivity and ongoing innovation exploded, creating unprecedented wealth and an ever-higher standard of living. Third, the failed attempt of Communism by China and the Soviet Union confirmed that capitalist Growth Economies were the preferred avenue to continued increases in human material well-being. The end of the Cold War and the opening of China added almost two billion people to the global capitalist economy.

Gross World Product Year 1 to 2015   Chart 5.1

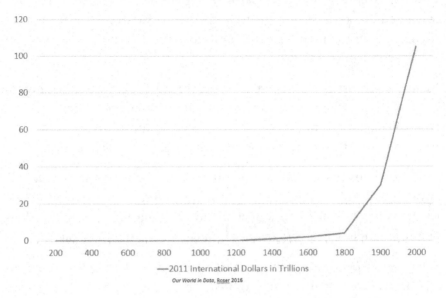

——2011 International Dollars in Trillions

Our World in Data, Roser 2016

This chart is startling. More wealth was created in the last 100 years than in all of human history. The Industrial Age delivered labor-saving technologies and spread new wealth around the globe. This incredible growth was fueled by fossil fuels – cheap energy that multiplied many times what each human could achieve without machinery. Electric lights lengthened our days. The 20th century also birthed the advertising industry and electronic media that together catalyzed unprecedented consumption.

As chart 5.2 shows, not only did total GDP grow, but per capita GDP grew by leaps and bounds. The vast amount of per capita growth occurred after 1950.

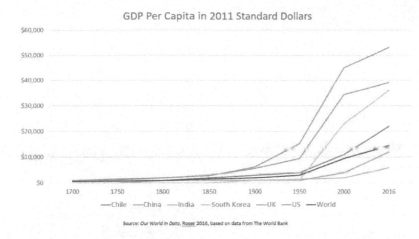

Source: *Our World in Data*, Roser 2016, based on data from The World Bank

Chart 5.2

Growth Economies exploded during the last 200 years. In the last 150 years, petroleum became the dominant fuel for humanity. 19th and 20th century inventions became institutionalized. Gasoline driven combustion engines became the standard mode of transportation. There was no reason to find another mode... until we learned that all those carbon emissions were harmful in many ways. The automotive industry and the fossil fuel industry were happily making money hand over fist, and had no viable competition, so had no reason to change.

Three dynamics transformed this scenario. The first is that, starting in 1950, there was a population explosion unprecedented in the history of mankind. The global population in 1950 was 2.5 billion and today it is 7.7 billion, a three-fold increase in just 70 years. To put that in perspective, it took 150,000 years for modern humanity to reach 1 billion people by the early 19th century. Every decade since 1950 had double digit population growth rates that only fell to single digits in the current decade.

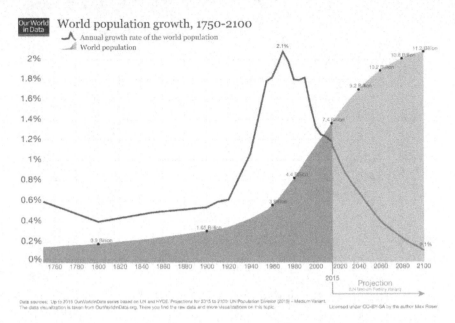

Chart 5.3

So, an unprecedented population explosion dramatically increased the use of fossil fuels and the amount of greenhouse gases being emitted globally. The long-term trend projected into the future shows population numbers plateauing, and eventually declining. The replacement rate for any population to maintain equilibrium is 2.1 children per woman of child-bearing age. Humanity reached that in the 1960s. We are now down trending from that peak and are currently at 1.4 children per woman of child-bearing age. In the age of climate change, lowering population growth is a major moral issue for all humanity.

The second dynamic was that, post-WWII, the United States (especially Hollywood and Madison Ave.) did a wonderful job marketing "the American Dream and the American Way of Life" around the world. The rest of the world saw the quality of life Americans enjoyed in our consumer culture, and they wanted to enjoy the same lifestyle. If everyone on Earth lived at the American standard of living, humanity would need five Earths worth of resources to maintain equilibrium. See Chart 4.2.

The third dynamic was the ever-ascending level of consumption in America. As recently as 50 years ago, American households had one car, one television, and one of each major appliance. Since then, families have evolved to a 'one car per person' paradigm. And every member of the family has a television or a computer, tablet or hand-held device (or all four). All of which require power, which is supplied largely by fossil fuels.

This ever-increasing consumption cut across all kinds of items. 50 years ago, it was rare to have a walk-in closet, now most new construction has at least one. In the past, the common man had no need for so much closet space. Today, many people do, as they fill closets with clothes made necessary by the perceived requirements of fashion. Consumption of stuff has increased geometrically since the mid-20th century.

The fourth and perhaps most significant dynamic has occurred in the last 30 years. While we were adding numbers to the total population of Earth, simultaneously people all over the world were moving up the socio-economic ladder. Hundreds of millions of people entered the middle class for the first time. These people had the wherewithal (and the access to credit) to consume at much higher levels than they had before. Since the early 1990s, the lead story of a nightly newscast could have been something like: "Around the world today more than 125,000 people have risen out of extreme poverty." This has happened every day for the last 25 years!

So, the last third of the 20th century saw a convergence of dynamics creating unprecedented and wide-spread material wealth, which dramatically increased consumption and resulted in adverse effects on planet Earth. The massive number of products produced, consumed and discarded globally created a severe garbage and waste problem, and pumped extraordinary amounts of greenhouse gases into the atmosphere.

It wasn't until recently that renewable energy became price competitive with fossil fuels. So, today it is possible to live affordably in a world powered totally by alternative and renewable energy. We don't have to keep burning fossil fuels if we choose not to.

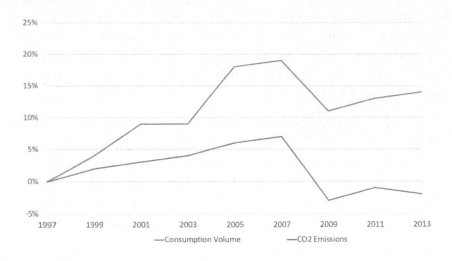

1997	1999	2001	2003	2005	2007	2009	2011	2013

——Consumption Volume ——CO2 Emissions

Chart 5.4 – U.S. Consumption and Emissions, Nature Communications 2015

Chart 5.4 illustrates how closely CO2 emissions follow economic output. When we experience a recession (in our Growth Economy), CO2 emissions immediately decrease. GDP, an output statistic, is purely a construct of Growth Economies. You can see here in chart 5.4 what happened during the "great recession" of 2007 to 2009. So, the decline in output and consumption was a "green event." That says it all… when Growth Economies are faring poorly, Earth benefits.

GDP will no longer be a valid measurement in a Finite Earth Economy. Ever growing GDP is counter to slowing global warming and overcoming climate change. In a Finite Earth Economy, GDP will give way to new economic metrics that include carbon footprints and offsets, and vastly lower consumption levels.

Major Takeaways

- Growth Economies have been the economic models throughout all history.

- Growth Economies, powered by fossil fuels and with an outcome of ever-increasing consumption, are the key drivers of global warming and climate change.

- The combination of unprecedented economic growth, wealth creation and population expansion has accelerated consumption and resulted in unacceptable levels of waste and greenhouse gas emissions.

- To save ourselves from ourselves, we must move away from Growth Economies.

- We can no longer live on a finite planet operating in an economic model that is based upon infinite growth. Our planetary spaceship has limited resources and we must work within them to avoid aborting the mission.

Chapter Six

The Circular Economy

"Reduce, Reuse, Recycle" – a popular phrase from Earth Day 1970

The Circular Economy is the vision of people interested in improving how humanity lives in harmony with nature. The "three Rs" are at the core of the concept of a circular economy.

Reduce consumption, reduce waste, reduce purchases, reduce the plunder of earth's resources. Reuse as much as possible. Find creative ways to reuse purchased products of all types. Recycle as much as possible. Buy products made with recycled materials. Buy products that have recyclable packaging.

This is correct thinking. Creating a circular economy minimizes waste and mindless consumption and creates a commitment to use less. A circular economy reduces the stress humanity places on Earth. A linear economy does not. Growth economies are largely linear: they take, make, sell, use, discard. The "discard" becomes waste, often after just a single use. Waste is intentionally built into the linear/growth economy. Planned obsolescence ensures we'll buy more and assumes Earth's resources are unlimited. Earth has finite resources.

Disposal often leads to the need to repurchase. Growth economies are measured on output. GDP is an output measurement. The growth of production and consumption is what linear/growth economies do. GDP does not measure economic efficiency. Growth economies are not efficient relative to Earth, circular economies are.

Here is Wikipedia's definition of a Circular Economy:

"A circular economy is an economic system aimed at minimizing waste and making the most of resources. This regenerative approach is in contrast to the traditional linear economy, which has a 'take, make, dispose' model of production. In a circular system resource input and waste, emission, and energy leakage are minimized by slowing, closing, and narrowing energy and material loops; this can be achieved through long-lasting design, maintenance, repair, reuse, remanufacturing, refurbishing, recycling, and upcycling.

Proponents of the circular economy suggest that a sustainable world does not mean a drop in the quality of life for consumers and can be achieved without loss of revenue or extra costs for manufacturers. The argument is that circular business models can be as profitable as linear models, allowing us to keep enjoying similar products and services.

To achieve models that are economically and environmentally sustainable, the circular economy focuses on areas such as design thinking, systems thinking, product life extension, and recycling."

The idea, vision and purpose of a circular economy are all well-meaning. In a time when we must consider what we take from Earth and what we give back, the concept of the Circular Economy is a reasonable response. An acknowledgement here to all who have been proponents and practitioners of the Circular Economy.

There is one big problem with the state of the Circular Economy in 2019: it only constitutes 9% of the global economy. It is the most efficient, planet-friendly segment of the global economy, but it is not significant enough to alter the apocalyptic situation we now find ourselves in. If only the Circular Economy had been more widely integrated after Earth Day 1970! Here we are 50 years later and only 9% of the global economy is Circular. A Circular Economy effective at stalling climate change would comprise 60% or more of today's global total.

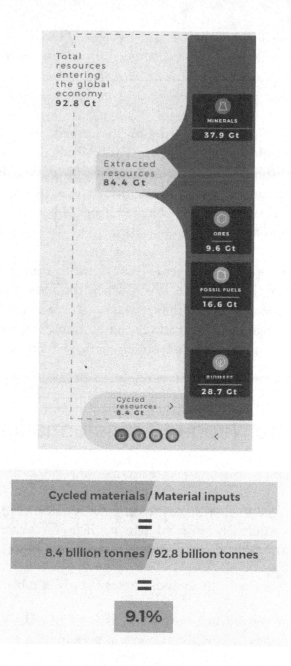

Total resources entering the global economy
92.8 Gt

Extracted resources
84.4 Gt

MINERALS
37.9 Gt

ORES
9.6 Gt

FOSSIL FUELS
16.6 Gt

BIOMASS
28.7 Gt

Cycled resources
8.4 Gt

Cycled materials / Material inputs

=

8.4 billion tonnes / 92.8 billion tonnes

=

9.1%

Charts 6.1 and 6.2 Source: 2019 Circular Gap Report
by the Circular Economy think thank

Part of the reason for this failure is that those pushing for a Circular Economy were fighting upstream, against vested interests, and a deeply entrenched mindset due to centuries of Growth Economies.

"There is nothing so difficult to plan, more uncertain of success, nor dangerous to manage than the creation of a new order of things. For the initiator has the enmity of all those who would profit by the preservation of the old institutions, and merely lukewarm defenders in those who would gain by the new ones." – Machiavelli

The leaders of the Circular Economy movement not only had the powerful strength of the installed global growth economy plus the carbon combustion complex working against them, they did not have the support of an intrinsic human motivator – the fight or flight response. We have that today. Climate change is here now, and most people view it as a clear and present danger.

As stated in Chapter 2, humans are not wired to react based upon a potential future threat. In the 1970s, when the movement first got traction, the threat of climate change was not "here and now" and was vulnerable to challenge. The carbon combustion complex could and did sow doubt; and framed it as something that would never happen.

The Three Needed Transformations

In Chapter 3 we discussed the three transformations needed by 2030:

1. Eliminate the burning of fossil fuels as quickly as possible, and significantly lower the rate at which we are consuming earth's resources.

2. Draw down the greenhouse gases already in the atmosphere.

3. Create "crew consciousness' in as many of today's 7.7 billion passengers on Spaceship Earth as possible.

The Circular Economy addresses part of #1 relative to lowering consumption, and thereby reducing the use of fossil fuels. The Circular Economy also relies on "crew consciousness" to a degree. If the Circular Economy had become a much larger part of the global economy, our current climate situation would not be as critical. The

Circular Economy does not address ending the burning of fossil fuels or the drawdown of GHGs already in our atmosphere.

Champions of the Circular Economy should be thanked for their efforts. Their vision was (and is) clear, but they lost the battle to gain global acceptance and implementation. Climate change is a real and present threat. A circular economy is no longer enough. A more radical and global step needs to be taken – a move to a Finite Earth Economy.

In moving to a Finite Earth Economy, we will adopt a Circular Economy reality along the way. It addresses the following issues:

- decreasing consumption and waste,

- designing products to last and to be fed back into the production cycle,

- ending the practices of disposability and fast fashion, and

- evolving past the conditioned response that something bought new is better than something bought used.

In both Circular and Finite Earth Economies, "used" is a positive term. Used means that something can be purchased without causing depletion of Finite Earth resources. A person who buys used (whether it's a car, a coat or a book) will gain social distinction, because how society and individuals perceive the concept of "used" will have changed.

Circular and Finite Earth Economies can be as profitable as Growth or linear economies. We need to disengage the concepts of profit, growth and efficiency as they have been defined in the Carbon Combustion Complex-fueled Growth Economy and redefine them for a future of living in a Finite Earth Economy.

The concept of a Circular Economy fits entirely inside the Finite Earth Economy. The behavior championed by Circular economists and proselytizers; can all be folded into the movement to a Finite Earth

Economy. It is a bridge of practice, opinion and advocacy that we must cross to successfully face the challenge of catastrophic climate change.

Takeaways

- The concept of a Circular Economy is a viable one and could have averted the catastrophic situation we now face. 9% adoption over a 50-year time frame is nowhere near enough and we are running out of time.

- The Circular Economy model fits into the Finite Earth Economy, wholly.

- The Three R's (Reduce, Reuse, Recycle) will continue to be key concepts in the Finite Earth Economy.

Chapter Seven

The Finite Earth Economy

Our Only Viable Way Forward

"If capitalism itself doesn't take into account natural capital, then we're all stuffed." – HRH the Prince of Wales

"You never change things by fighting the existing reality. To change something, build a new model that makes the existing model obsolete." – R. Buckminster Fuller

The Growth Economies we have all lived in throughout history, now powered by fossil fuels, have delivered us to the existential crisis we are now facing. Even if these growth economies had converted to alternative sources of energy a decade ago, the unlimited consumption model they represent would still be stressing ecosystems everywhere.

Humans are the only species that creates "waste". The waste that is fundamentally built into Growth Economies is a major problem for us and all our fellow species.

This chart from Chapter 3 frames all three economies discussed in this book:

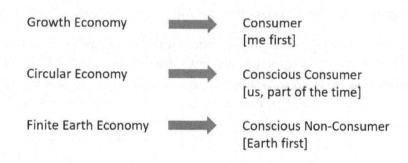

The Finite Earth Economy (FEE) places Earth first. That is the call to action for moving to a FEE: Earth First!

What does "Earth First!" mean?

It means that economic decisions at all levels – personal, local, state, nation, global – are made with the health of Earth first in mind.

Use of any earth resource: water, air, minerals, plants and animals must be captured inside corporate balance sheets. This must also occur at the personal, neighborhood and local levels. We can no longer discount the Earth's bounty as if it is free for the taking. The same is true for any ecological damages a business might cause. For example, an oil spill would require whatever it takes to restore the natural world (land, water, flora, fauna) back to its pre-spill conditions. And the company responsible for the spill will be 100% responsible for paying for the cleanup and restoration.

Tax laws and codes at all levels of government will focus on the carbon footprint of all entities. For example, a corporation that is carbon neutral will receive a significant tax deduction. A corporation that uses fossil fuels for less than 50% of its energy will receive a lesser deduction. A corporation that uses fossil fuels for 100% of its energy will have a tax penalty. If a corporation reduces its carbon footprint at least 10% from year to year, they will receive a tax incentive. These new carbon footprint-related taxes will also apply to individuals, families and businesses of all sizes.

Any entity that promotes itself as being green must be 100% carbon neutral, or they will be fined. Consumers will be informed that only companies that are 100% carbon neutral can market themselves or their products with a green message. This will be part of a complete realignment of the advertising and marketing businesses.

A large, efficient carbon offset marketplace (perhaps overseen and coordinated by governments and nonprofits) will be established. A full Carbon Capture Crew, similar to the Civilian Conservation Corps (CCC) of the 1930s, could be deployed in every country. Funded by carbon footprint taxes, their tasks might include planting millions of trees, painting roofs white to reflect more sunlight, paving roads with recycled plastics, installing carbon capture technologies, etc.

An energy policy that is 100% designed to support all non-fossil fuel energy sources will be implemented. Allocated funds will be used for research, development, deployment and anything else that might accelerate the transition to renewables and alternatives as quickly as possible. The first step would be to move the $5.3 trillion of worldwide annual government subsidies[1] away from fossil fuels and toward energy sources that emit no GHGs.

At all levels, there will be major initiatives to educate the public on the urgency of the situation, and to make them ever more conscious non-consumers. The advertising and marketing industries will be enlisted to make this their new endeavor. The overall goal being to change people's habits... evolving them from passive consumers to Conscious Non-Consumers. First awareness, then status, then mainstream — resulting in a new Finite Earth Economy business as usual.

The label "used" will gain cachet. Instead of having a negative undertone, "used" becomes something to be proud of, something that bestows social capital, something that the average person seeks out.

It would replace fast fashion with slow fashion. Instead of new "styles" every year, it will be fashionable to buy or make the most durable, high quality clothing and other products. The best designed pieces will be repairable, and at their end of life will be easily fed back into the production cycle to create a new piece.

Wherever possible, it would transition physical products into virtual ones. This has been done with content. There's no need to buy CDs, DVDs, books and magazines in their physical form. This means three things. First, in homes, shelves of stuff are replaced by a small 2 Terabyte hard drive. Second, there is minimal CO_2 generated in production. Third, there is no CO_2 generated in the distribution of digital content, except the electricity of the devices used.

There are several macro-trends that are currently happening that align with the Finite Earth Economy:

The Sharing Economy – Share what you own and what others own. Share almost anything such as cars, extra rooms, tools, books, kitchen utensils... anything that is only used occasionally. The average American uses their car 4% of the time, the rest of the time it is just sitting there depreciating. Why own a car? Why own a hammer when you need to use one three times a year? Community centers can house and manage the inventory, and share what others own like a lending library.

Moving from Ownership to Rental – This has already happened with content streaming services. And it's happening with transportation (cars, scooters and bikes) and housing. How many other items might we rent for occasional use instead of buying them?

The Sufficiency Movement – This is a spiritual and consciousness concept being adopted by many. It's a knowing that there is enough for all and there's no need to scramble and fight. It's the difference between collaborating on how to allocate resources so that all are served and acquiring as much as possible by any means possible.[2] It's a true practice of Spaceship Earth Crew Consciousness.

"There is a sufficiency in the world for man's need but not for man's greed." – Mahatma Gandhi

Simplifying – Is living happily with less. There's a freedom and stress reduction that comes with downsizing… with eliminating the stuff we don't need and creating space. Marie Kondo is a popular thought leader of this trend. She, and many others, have made successful careers out of teaching people to simplify.

"The ability to simplify means the ability to eliminate the unnecessary so that the necessary may speak." – Hans Hoffman

Preferring Used over New – Other than food, beverages, personal products and medicines, think used first. This will ultimately creep into the social mores of societies. Major purchases of used items will be embraced and admired. Buying new will be questioned, not respected.

The above are just a few samples of what a Finite Earth Economy will entail. It not only can be done, we must do it. We must save ourselves from ourselves.

In Book 2 we will address the fundamental metrics that must change, the first major economic actions to take, new timelines and achievable milestones, and the essential technologies needed to achieve the goals stated in Chapter 3 and the Manifesto.

In Book 3 we will address the consciousness, actions, policies, laws and behaviors needed at the personal, local, state, national and global levels. We will underscore why the mantra for all of this is:

URGENCY/URGENCY/URGENCY

Potential Paths Forward in the Age of Climate Change

Now in 2019 there are three loose paths to 2100. They are: Catastrophic, Risky and Best Probable. Only the last one gives us a fighting chance for a viable human civilization by the end of this century.

Catastrophic

Business as usual with a gradual move away from fossil fuels over time is no longer an option. Current forecasts, if we don't take drastic action now, put the rise in temperature at 3 to 7 degrees Celsius. We largely continue to stay on the course we are on while gradually implementing a move away from fossil fuels over time. An incremental global transition to renewables that is slower than needed with a focus on getting it right by 2100. The temperature rise here is hard to predict, but current forecasts put the rise at 3-7 degrees C by 2100. It would be a planetary environment never experienced by humanity.

Civilization may collapse, but complete extinction isn't likely. Large swaths of the planet will be uninhabitable by humans. Our descendants, those who survive, will reside in a nightmare world that we can't even imagine.

"Most of the planetary surface would be functionally uninhabitable. Agriculture would cease to exist everywhere, apart for the polar and sub-polar regions. The oceans would stratify and become oxygen-deficient, which would cause a mass extinction. It's pretty much equivalent to a massive meteorite striking the planet, in terms of the overall impacts." – Mark Lynas, "Six Degrees: Our Future on a Hotter Planet"

Risky

We focus our energies on demonstrations, protests, and the signing of Accords as meaningful action. Slowly chipping away at the Carbon Combustion Complex, or assuming market forces will fix the problem, won't work. This approach references guidelines to be in place by 2050. An easy number to embrace as we can give ourselves a long lead time to make fundamental changes. We are happy with small victories, a gradual change from fossil fuels to alternative and renewable energy sources and looking at a 30-year transition period.

This approach is too slow and takes too long. We almost assuredly will pass more tipping points that trigger runaway global warming that will take a century or more to constrain. The result will be massive suffering for man and most other species.

"We often preoccupy ourselves with symptoms, whereas if we went to the root cause of the problem, we would be able to overcome the problem once and for all." — Wangari Maathai

Best Probable Future

We must move to a Finite Earth Economy as fast and completely as possible.

We must set a deadline of 2030 for fundamentally changing how humanity powers its civilization. We must begin extreme drawdown efforts for eliminating GHG emissions and removing resident CO2 from the atmosphere.

We must evolve toward a crew consciousness in how we relate to and interact with Spaceship Earth.

Our future depends upon urgent mobilization and a deadline of 2030 to make these fundamental changes. We move forward with a vision of massive and rapid change, an ever-growing critical mass of humanity, all embracing this necessary change in how we live.

We already have triggered something (more likely multiple things), that has caused climate change to happen sooner than predicted. We must act now to prevent any additional tipping points from occurring.

Humanity must acknowledge and take full responsibility for global warming, and take urgent and wide-ranging action to slow it down by 2030.

We accept the current reality of a 1degree Celsius warming of the surface of the planet and strive to keep that below 1.5 degrees. This path allows humanity to enjoy a thriving, evolved civilization by the end of the Earth Century.

How can we possibly move a significant percentage of humanity to a Finite Earth Economy by 2030? It will take vision, clarity, will, persistence, forward facing investments and massive collective action, all infused with a deep sense of urgency.

"If the success or failure of this planet and of human beings depended on how I am and what I do... How would I be? What would I do?" – R. Buckminster Fuller

We are ready. Are you? Are you ready to make huge changes in your life and your business for the sake of humanity and countless other species on the planet?

Then move on to Book 2 and Book 3 so you can do your part to institute a Finite Earth Economy as soon as possible.

Moving to a
Finite Earth Economy

Crew Manual

Book Two

New Metrics and Technologies

Chapter Eight

Fundamental Metrics Must Change

The secret of change is to focus all of your energy, not on fighting the old, but on building the new." – Socrates

"There is nothing so difficult to plan, more uncertain of success, nor dangerous to manage than the creation of a new order of things. For the initiator has the enmity of all those who would profit by the preservation of the old institutions, and merely lukewarm defenders in those who would gain by the new ones." – Machiavelli

It is perhaps the most difficult of endeavors, to envision and then create a new way to organize how humanity functions. This is even more difficult when what needs to change is the global economy itself. Everyone reading this has grown up in some form of a Growth Economy. It is all we know. It is the basis of most economic theory. It is our way of life.

This economic way of life, as discussed in Book 1, threatens the existence of humanity and most other species living on Spaceship Earth. Growth Economies, which require ever more production and consumption and are powered by fossil fuels, are the key causes of global warming and climate change.

Potential Paths Forward in the Age of Climate Change

In Chapter 7 of Book 1, we wrote about three potential paths forward to 2100. They are: Catastrophic, Risky and Best Probable.

The Catastrophic path is business as usual with a gradual move away from fossil fuels over time. Current forecasts, if we don't take drastic action now, put the rise in temperature at three to seven degrees Celsius by 2100. Life as we know it today would be a fond memory.

The Risky path focuses our energies on demonstrations, protests, and the signing of Accords as meaningful action. Slowly chipping away at the Carbon Combustion Complex is too little too late as greenhouse gases continue to be emitted and continue to accumulate in the atmosphere. The result will be massive suffering for man and most other species.

To all activists who read this, please reread the quote from Socrates at the top of this chapter. Everything that you are doing, that we are doing to raise consciousness and demand change IS bringing about change. We have reached a critical mass of protest and awareness. This coincides with what we wrote in Book 1, that climate change is NOW and it is triggering our human fight or flight reaction. So we must now "focus our energy… on building the new". To now move on to build the global Finite Earth Economy, which is the Best Probable path to a livable 2100.

The Best Probable path requires a transition to a Finite Earth Economy, and by 2030, gives us a fighting chance for a viable human civilization at the end of this century. The stakes are the preservation and continued evolution of humanity versus trying to preserve the status quo... we have no choice. It won't be cheap or easy, but there are precedents.

In the Addendum we have placed a number of links to times in the past when a common enemy or a major undertaking rallied us to alter the game to win, to commit vast resources to do so. These include the fact that, in today's dollars, WWII cost $4 trillion spent over four years. That the ramp up to this war effort totally mobilized the U.S. and cut unemployment by 60%. The return on investment of the hundreds of billions of dollars spent on the Apollo project and the entire space program in the 60s and 70s was 700%. That's a seven fold payback on the financial investment.[1]

In Chapter 10 we will list the new metrics and in Chapter 12 we will place them along a timeline. This is a timeline we must try to stay with if we want our children and grandchildren to have an evolving civilization rather than one in a downward spiral of societal collapse and eventual extinction.

We must move away from Growth Economies! That is the only way our offspring and their offspring will be able to evolve and live in a dynamically uplifting civilization.

Here are the metrics of Growth Economies. They are used to measure the well-being of nation states, to the exclusion of almost any other measurement. When looking at these as an individual, it becomes clear that they measure us only as consumers: not citizens, humans, sentient beings... just consumers.

There are two types of indicators you need to be aware of:

1. **Leading indicators** often change prior to large economic adjustments and, as such, can be used to predict future trends.

2. **Lagging indicators**, however, reflect the economy's historical performance and changes to these are only identifiable after an economic trend or pattern has already been established.

Leading Indicators[2]

1. Stock Market

2. Manufacturing Activity

3. Inventory Levels

4. Retail Sales

5. Building Permits

6. Housing Market

7. Level of New Business Startups

Lagging Indicators[2]

1. Changes in the Gross Domestic Product (GDP)

2. Income and Wages

3. Unemployment Rate

4. Consumer Price Index (Inflation)

5. Currency Strength

6. Interest Rates

7. Corporate Profits

8. Balance of Trade

9. Value of Commodity Substitutes to U.S. Dollar

(Gold and silver are often viewed as substitutes to the U.S. dollar.)

Take a look at this list. Where is the measurement of well-being of the citizenry? Where is the measurement of the well-being of Earth? Where is the measurement of sustainability through stewardship of the planet? Where is the measurement of anything related to pollution, waste, health of the lands and oceans? Growth Economy metrics reward behaviors that lead to pollution, poor health, depression and destruction of the biosphere's ability to support life.

It is estimated that some 150 species are becoming extinct every day! Depending on the source, this is 1,000 to 10,000 times the historical extinction rate.[3] How humanity lives on Spaceship Earth, living in Growth Economies based on consumption and powered by fossil fuels, is the trigger of the Sixth Extinction event in the history of the planet.

Growth Economies and the metrics that measured them have created historically unprecedented wealth, higher standards of living, scientific breakthroughs, and just about all non-natural elements of civilization. They served their purpose, but they are no longer productive. Too much of a good thing, they are now literally killing us and all living things. Growth Economies have run their course. Their metrics must be retired. Now, in the first quarter of the 21st century – the Earth Century – we must move to Finite Earth Economies. With urgency!

Takeaways

1. There are three paths to the end of the 21st Century: Catastrophic, Risky, and Best Probable. Humanity clearly must move from Growth Economies to a Finite Earth Economy which is the Best Probable path.

2. Metrics must therefore change from measuring only economic growth, to metrics that focus on well-being and the health of the biosphere.

Chapter Nine

First Major Economic Actions

"The journey of a thousand miles begins with one step." — Lao Tzu

Before we list the major economic actions we must immediately implement, we should take a look at where these actions will have the greatest effect toward cutting fossil fuels to 30% by 2030.

The transition from Growth Economies to Finite Earth Economies will focus largely on the top 20 economies by 2030. This will be a long and difficult process. Some countries are already on the path and will move more quickly. These are mostly smaller, more homogenous nation states, for example those in Scandinavia. The top ten countries based on the GDP metric must obviously take the lead as they represent two thirds of the global GDP and fossil fuel use.[1]

Here are two charts. The first ranks the top ten countries by GDP. The second chart is a proportional chart of all national GDPs. (Please keep in mind that with both economic and environmental statistics there seems to always be some variance depending on sources.)

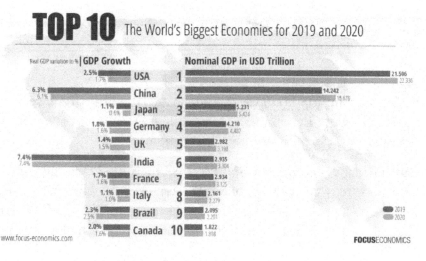

Chart 9.1

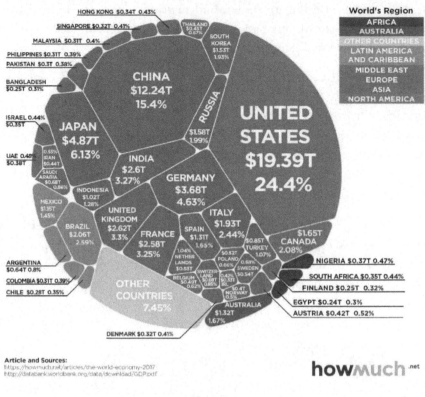

Chart 9.2

We look at the top countries based on GDP as they represent the largest Growth Economies. Some of the largest countries by population have significantly smaller economies.[2]

Remember the goal is to be at 30% fossil fuels by 2030 – globally. If these top ten Growth Economies, can exceed the 70% clean energy/30% fossil fuels level, then we will be on track to goal. Dozens of other countries can approach 70% and we will make our goal globally.

There are several points to make with these charts.

First, the United States has the single largest GDP in the world. It must lead the way as it represents the greatest opportunity to reach 70% clean energy by 2030. The United States is also, historically, responsible for persuading the world that Growth Economies are a good thing. Post WWII, the United States was the country whose advertising and entertainment flooded the rest of the world depicting the consumer society as the desired economic model. This was partly due to the Cold War as the United States was touting Capitalism in opposition to the Soviet Union touting Communism. We Americans led the world to the catastrophic level of consumption and fossil fuel use that exists today. We must now lead the world to a Finite Earth Economy. It is our responsibility and our karma if we want to lead the world to this new global economic order. We must lead.

Second, if you look at the top ten countries by GDP you will see that all except India – and still large parts of western China – are relatively wealthy. The wealthy countries must lead the way. The poor countries do not have the resources for rapid transformation. Additionally, there is growing evidence that the poorer countries will disproportionately suffer the adverse consequences of climate change. There is a moral imperative here.

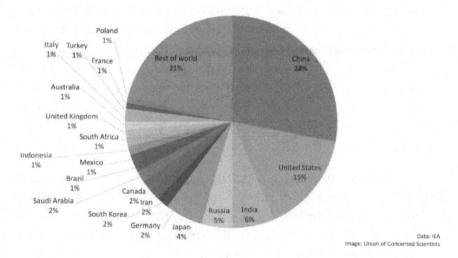

Chart 9.3

The chart above lists the top 20 polluting nations. There is a good deal of overlap between the list by GDP and the list by level of GHG emissions. In moving to a Finite Earth Economy, it is extremely important to have all the countries on these two lists be fully accountable for their emissions. Focusing on these countries will have disproportionate effect. Poorer, smaller countries have done the least damage and they have the least economic resiliency. Actions taken by the major polluters and largest/most resilient economies can bring the world to 70% clean energy by 2030.

First Major Economic Actions to Take

What follows is not a chronological list. It is a list of all the major actions we must start now. Some will move more quickly than others. Some will, as the Machiavelli quote suggests, have strong resistance from the status quo and the legacy thinking behind it. We must start now on all of these enormous undertakings.

Carbon Fee and Dividend

The implementation of a Carbon Fee and Dividend policy is the most important step to take in every major economy. The difference between "fee" and "tax" will be explained shortly. Placing a fee on carbon simply means that we are putting a "price" on carbon above its historical economic cost. Remember that there are significant costs to society with carbon: damage to the environment, and serious human health costs. What price can be put on this, we cannot state exactly, but it is in the trillions annually.

So we increase the cost of carbon and that will drive down consumption. There are many precedents to this in Growth Economies. Taxing cigarettes is perhaps the most recent and apropos example. When the initial reports that smoking cigarettes caused cancer were published, there was no discernible drop in smoking. That only occurred over several decades because tobacco was taxed... heavily. That made the difference. In the United States the rate of smoking dropped 67% from 1970 to 2015.[3]

When seat belt laws went into effect in the U.S. highway deaths plummeted. Even though there was general knowledge that wearing seat belts was a much safer way to drive, the independent "it won't happen to me" attitude persisted. Only after the laws were passed did buckling up become a societal norm.

Fee and Dividend is based on the Free Market Economics of Milton Friedman. Place a fee on C02 by the ton and then redistribute the money to consumers to offset the higher cost of carbon at the gas pump and elsewhere. An initial plan, submitted to Congress, set forth by top Republican advisors to Ford, Reagan and George H. W. Bush suggests that in the U.S. the revenue from a $40/ton carbon fee would generate $3 billion annually. This would be distributed to consumers at an estimated $2,000 per family of four.[4]

There are many benefits to this Fee and Dividend plan.

- It will immediately increase the cost of carbon, resulting in a decrease in fossil fuel use and emissions.

- If it is fully implemented, many emission regulations can be relaxed or removed since the end result will be the same. So conservatives should support Fee and Dividend. At the same time, environmentalists should fully support putting a price on carbon as it enters the economy.

- Receiving a $2,000 check for a family of four will be welcomed. Consumers/voters will support an increase to $50/60/70 a ton because their annual dividend checks would increase accordingly. This is why it must be a Fee and not a Tax, so the money goes back to the consumers and voters, not government coffers.

- This is a wealth redistribution from the largest emitters to the individual citizens that aligns all with the end goal of zero emissions.

National Carbon Footprint Mechanisms and Measurements

As we will address in Book 3, in a Finite Earth Economy a large part of national tax codes will be based upon carbon footprints. It is therefore imperative to develop accurate, scientifically agreed upon ways to measure so that all corporations, governments, nonprofits, educational institutions, places of worship and individuals can learn what their footprint is. They then will be motivated to learn how to lower it. The requisite training will be provided. It will become a social norm to know your footprint and "keeping up with the Joneses" will take on a new meaning. Every organization of every size, from individuals to the largest multinational corporations will be rewarded or penalized on their carbon footprints.

Education

Education concerning how the natural world works (how everything is interconnected into one system) will start early. School children will learn how much CO_2 is emitted to prepare and deliver their Happy Meals. The list of concepts that all citizens should understand will include the cost to nature of agriculture, manufacturing, transportation. Education on all aspects of the Circular Economy and the deadly costs of carbon will be universal.

Appropriate curricula will be developed for all grades from K-12, to higher education to advanced degrees in climate science, renewable energy and more.

Massive Changes to Agriculture

This is a big one and must be started ASAP as it will take time to redesign the current high GHG emissions model without disrupting food production. More on this in Chapter 11.

National Service to Clean-up, Conserve, Create Carbon Offsets

A national call to action for creating carbon offsets overseen perhaps by nonprofits and the government. Governments will allocate some of the carbon footprint taxes to set up a national agency, staffed by climate scientists, activists and even entrepreneurs. This has the elements within it of the New Deal of the 1930s, which has been partially set forth by the recent Green New Deal, but with different dynamics and sources of funding. Those that pollute the most will be the largest funders via taxes.

The Civilian Conservation Corps (CCC) was created in 1933 by FDR to combat unemployment. The CCC was responsible for building many public works projects that are still in use today. This is particularly relevant to today as there is so much work to be done in every country: planting trees, restoring habitats destroyed by pollution and toxic chemicals, cleaning up plastic waste, etc.[5]

These projects will put millions of unemployed and underemployed people to work. If these projects were folded into a mandatory two-year national service with the initial purpose of working on all things that will result in habitat restoration and emissions reductions, there would be an alignment that would not only train young people, but would create a society that values cooperation. This will be discussed in more detail in Book 3.

Transportation

By 2030, 50% of all cars in the U.S. will be electric vehicles. National and local governments will subsidize this transition by buying any vehicle that gets less than 20 mpg and is driven more than 10,000 mile per year. Incentives will be granted for electric cars and for autonomous cars with multiple riders.

Subsidize, and therefore accelerate, the movement to ever better battery technology for air travel, municipal buses and trains. Incentivize private industry to support this needed and massive undertaking in every country.

Takeaways

1. The largest GDP countries and the countries with the greatest GHG emissions must take the greatest responsibility for leading the world to a Finite Earth Economy.

2. Smaller, but wealthy countries (e.g Switzerland) must also lead to give poorer countries without the wherewithal more time to make the transition.

3. Nations must implement Carbon Fee and Dividend laws ASAP.

4. In preparation for a carbon based tax code, countries must set up clear, transparent and measurable guidelines for carbon footprints.

5. Agriculture, Education, and Transportation sectors must all immediately start their move to a Finite Earth Economy. Details are in Chapter 11 and in Book 3.

6. The U.S. and the top 10 GDP countries and top 10 wealthiest countries must set up national service programs to clean-up, conserve, retrofit and create carbon offsets. This should also be done in poorer countries with high unemployment.

Chapter Ten

The New Metrics

"If you don't collect any metrics, you're flying blind. If you collect and focus on too many, they may be obstructing your field of view." – Scott M. Graffius, Agile Scrum

"We should use Nature as the measure, using Nature's wisdom as a template for our economic systems." – Douglas Tompkins

"Tell me how you measure me, and I will tell you how I will behave." – Eliyahu M. Goldratt

"What can be counted doesn't always count, and not everything that counts can be counted." – Albert Einstein

"People with targets and jobs dependent upon meeting them will probably meet the targets – even if they have to destroy the enterprise to do it." – W. Edwards Deming

Obviously, the metrics of the Growth Economies in which we all live must change for the move to a Finite Earth Economy. We must create new metrics to measure the health of the Finite Earth Economy. During the last 100 years the "health" of Growth Economies has been the primary measurement of a nation state's well- being. If a country's economy is expanding, then the country is in a good state. Beyond the obvious problem of economic expansion as the top measurement of a country's well-being, there is now the catastrophic problem of a climate crisis created and intensified by a perceived need for continuous growth.

Gross domestic product (GDP) is the most used measurement of a nation's economic health. It measures the output of the linear Growth Economy. It is largely about output, not efficiency, nor waste or pollution… and certainly does not put the earth first. GDP will still be a fair measurement, but will no longer be the most important. Ironically, climate change is the single largest threat to GDP growth. Climate change is now reducing global GDP by at least 1% a year, and its impact will only grow going forward.[1]

As of this writing, the annual global GDP is around $90 trillion. This means that there will be a minimum of $10 trillion dollars deducted from global GDP by 2030. Reductions include property damage, business losses and trade interruptions. If we do not move to a Finite Earth Economy as quickly as possible, the percentage of GDP lost to climate change will increase exponentially in the 2030s. Climate change is a more serious risk to the global economy than anything else. The World Economic Forum has been suggesting this for the past several years. In the chart below #1, #2, #6, #7, #8 and #9 are climate-related.

1	Extreme weather events (e.g. floods, storms, etc.)	
	Failure of climate-change mitigation and adaptation	2
3	Major natural disasters (e.g. earthquake, tsunami, volcanic eruption, geomagnetic storms)	
	Massive incident of data fraud/theft	4
5	Large-scale cyberattacks	
	Man-made environmental damage and disasters (e.g. oil spills, radioactive contamination, etc.)	6
7	Large-scale involuntary migration	
	Major biodiversity loss and ecosystem collapse (terrestrial or marine)	8
9	Water crises	
	Asset bubbles in a major economy	10

Top
10
Risks by
Likelihood
Global Risks Report

Chart 10.1 Source: World Economic Forum[2]

It is clear that pure growth for its own sake needs to end as soon as possible. It is also clear that growth does not always translate to profit. While the essential need to move to a Finite Earth Economy is clear, there will be a disruptive transition. During this transition we will have to work to determine how capitalism and free enterprise can remain viable.

Moving to a Finite Earth Economy is NOT about ending Capitalism. It is recasting it with an 'earth first' and 'humanity first' focus. It will be a rocky transition that won't end until we can stabilize global warming and reshape our Capitalist reality. We think this can be done by 2030… if we develop the will to do so.

Capitalism, a word largely shaped linguistically in the 19th century, was a form of economic theory that put the economy in private hands. This was in contrast to Socialism, a form of economics where the state had control of and responsibility for the economy. Today, most countries in the world have developed economies that have some form of socialism. In the U.S. we have public education, Social Security, Medicare, Medicaid, the management of national parks and, of course, national defense. We have accepted these services as part of the American way of life. We will need to develop new governmental programs, in coordination with the private sector and nonprofits, that must be planned, designed and implemented with urgency.

The range of estimates relative to the global economic damage that climate change will cause is a wide one. To state what I think is an approximate number: every day that humanity is not mobilizing to move to a Finite Earth Economy is one billion dollars in future damages. That means that in 2019, as emissions and consumption increase, we will be adding 365 billion dollars of damage in the future. This should be thought of as a minimal cost of delay. That is why urgency is absolutely key. The faster we move to a Finite Earth Economy, the lower the ultimate economic cost will be.

We must move to new metrics to measure the efficiency of the Finite Earth Economy and its health going forward.

Possible New Metrics

The Social Cost of Carbon

The social cost of carbon reflects the global damage of emitting one ton of carbon dioxide into the atmosphere, accounting for its impact in the form of warming temperatures, more severe storms, rising sea levels, etc. Economists see it as a powerful policy tool that could bring rationality to climate decisions. It's what we should be willing to pay to avoid emitting that one more ton of carbon. In real ways, this is the truest and most important metric relative to living in a carbon fueled economy.[3]

Unfortunately, the true dollar number assigned by economists is hotly debated and it ranges from $40 per ton of CO2 to $400 per ton. This extreme variation is due to what goes into the calculations… what should be included and how it's measured. Recent computations suggest that mortality alone is $30 a ton. Someone dies every 6 seconds on earth due to a health complication from air pollution. Then there are other costs such as changing weather patterns. How much does a prolonged drought reduce farm revenue? What are the costs of more frequent and severe flooding? What is the cost for a region to have a "once in 500-year storm" every two or three years?

In a world where 90% of humanity is breathing polluted air (which kills seven million people every year), the health effects of air pollution are serious – one third of deaths from stroke, lung cancer and heart disease are due to air pollution. This is an equivalent effect to that of smoking tobacco. The social cost is enormous.[4]

The social cost of carbon (whatever the dollar number) will disappear with the move to a Finite Earth Economy. Eliminating carbon emissions will be the single best public health effort we can make. Moving to a Finite Earth Economy will dramatically improve health for all living things.[5]

Moving Earth Overshoot Day Back

This metric addresses levels of consumption and waste.

In Book 1 we discussed how Earth Overshoot Day (the date in the current year when humanity has used 100% of the resources the planet can regenerate) is now in early August. Ideally it should be after midnight on December 31. If it were, we would be extracting and consuming only as much as the Earth can regenerate. Since 1972 we have not lived within that constraint. We now consume the equivalent of 1.7 Earths every year. Obviously, this cannot continue. In 2018 Earth Overshoot Day was August 1.

By country there is a wide range of national Earth Overshoot Days. Honduras is the only country that is at 12/31. Luxembourg is the earliest at 2/17. Australia, Canada and the U.S. are close together at 3/12, 3/13 and 3/14 respectively. Much of Europe is May and June. China is 6/23 and Brazil is 7/26. The bigger a country's economy and the earlier its Overshoot date, the more that country is living beyond its resource means. Those countries must take the most drastic and immediate action.[6]

Another metric then, is to move Earth Overshoot day back a few days per year, with the goal to move as much as possible to December 31 by 2030. So we should set an annual target for Earth Overshoot Day.

Here is a suggested timeline for moving Earth Overshoot Day back:

Earth Overshoot Goals

2020	August 4
2021	August 8
2022	August 14
2023	August 20
2024	August 28
2025	September 10
2026	September 22
2027	October 5 *
2028	October 20
2029	November 5
2030	November 30

*Same level as 1995

This metric will show how well we are doing in reducing our consumption and waste. It is this metric that will place strong downward pressure on GDP. To no longer regard declining GDP as a negative phenomenon, we need to evolve to regarding the movement of Earth Overshoot Day back every year as a sign of success.

Percentage of Total Energy that is Fossil Fuels

The rapid conversion away from fossil fuels to all forms of clean, non-polluting energy is crucial. Currently the percentage of global energy that is obtained from fossil fuels is 77%.[7] This number is now measured annually.

The 2030 goal is no more than 30% of global energy provided by fossil fuels. On a flat average that works out to about 4.25% reduction per year for 11 years. The reality is smaller decreases will happen early in the decade, and larger decreases will occur later in the 2020s as clean energy is widely deployed. At the same time, the total amount of power used globally will grow along with the population and increasing quality of life.

Target Fossil Fuel Percentages of Total Energy Consumption

2019	77%
2020	76%
2021	74%
2022	72%
2023	69%
2024	65%
2025	60%
2026	54%
2027	47%
2028	39%
2029	30%

Instead of waiting for year end to tally our progress, we will be able to measure fossil fuel use on a daily basis. We cover the various technologies that will enable this in Chapter 11.

Global and National Annual Totals of Greenhouse Gas Emissions

Currently humanity is emitting approximately 37 gigatons of greenhouse gases each year. As mentioned in Book 1, that approximates the aggregate weight of discharging 315 four-ton African Elephants into the atmosphere every second. If CO2 was visible, we would have been shocked into addressing the issue years ago. Imagine leaving your home each day and seeing more and more elephants floating in the sky. You would know something was terribly wrong. However, GHG emissions are invisible, so we don't see the problem or the fact that it is growing every day. We have trouble believing what we can't see. This is an underlying reason for resistance to acknowledging climate change. GHG emissions are invisible. The several millimeter rise in sea level each year is hard to see.

Globally, the metric should be an average reduction of 5% of GHG emissions per year over 11 years. There are two issues with this metric. First, in the beginning, 5% will be impossible to reach because the requisite solutions and technologies will not have been deployed. Second, as the resident amount declines, 5% reductions will translate to lower total tonnage. These mean that we have to aim at actual total emissions per year as a goal. Here is a suggested global guideline using an estimated 2019 number of 37 gigatons of emissions:

Annual Global GHG Emissions

2020	37 gigatons
2021	36 gigatons
2022	34 gigatons
2023	32 gigatons
2025	29 gigatons
2026	25 gigatons
2027	21 gigatons
2028	17 gigatons

2029	13 gigatons
2030	10 gigatons

The underlying assumptions are that we will start slowly and then accelerate through mid-decade and reach a low point of maintenance in 2030 that can decrease by a constant 5% until 2040.

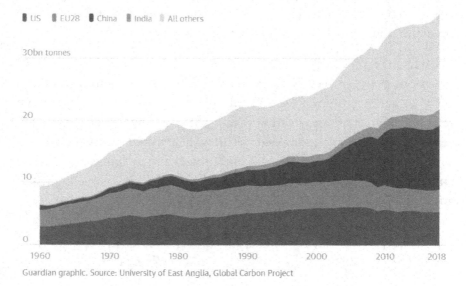

Global carbon emissions in 2018 are set to hit an all-time high of 37.1bn tonnes

US EU28 China India All others

Guardian graphic. Source: University of East Anglia, Global Carbon Project

Chart 10.2

Share of Global CO2 Emissions by Country in 2015

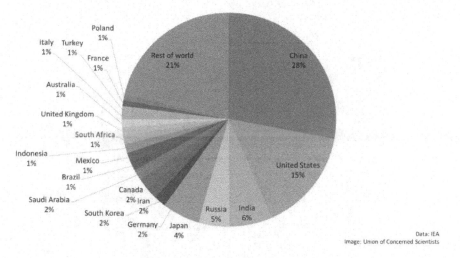

Poland 1%
Italy 1% Turkey 1%
France 1%
Australia 1%
United Kingdom 1%
South Africa 1%
Indonesia 1%
Mexico 1%
Brazil 1%
Canada 2%
Saudi Arabia 2%
South Korea 2%
Iran 2%
Germany 2%
Japan 4%
Russia 5%
India 6%
United States 15%
China 28%
Rest of world 21%

Data: IEA
Image: Union of Concerned Scientists

Chart 10.3

The top 20 emitting countries comprise 79% of total emissions. The remaining countries, of which there are 175, are responsible for only 21%. That's an average of 0.12% of annual emissions per country.

To make this an even starker comparison, cumulative emissions (how much each country put into the atmosphere since the Industrial Revolution) place the U.S. first at 25%, the European Union second at 22% and China third at 13%.

It is essential that these top 20 emitting countries lead the way. If the goal is to limit energy from fossil fuels to 30% by 2030, these top 20 countries, currently responsible for 79% of emissions, could get to 10% and the other 175 countries would not have to do anything at all. At the very least, the United States, China and the Eurozone must bear the largest burden as they are the largest three producers of greenhouse gases in the world – currently and historically.

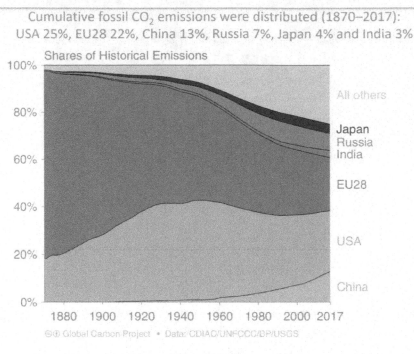

Cumulative fossil CO$_2$ emissions were distributed (1870–2017): USA 25%, EU28 22%, China 13%, Russia 7%, Japan 4% and India 3%

Chart 10.4

Per capita emissions (per person) by country sheds even more light on the story:

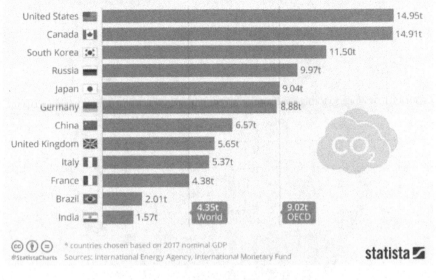

The Global Disparity in Carbon Footprints
Per capita CO_2 emissions in the world's largest economies in 2016* (in metric tons)

Country	Emissions
United States	14.95t
Canada	14.91t
South Korea	11.50t
Russia	9.97t
Japan	9.04t
Germany	8.88t
China	6.57t
United Kingdom	5.65t
Italy	5.37t
France	4.38t
Brazil	2.01t
India	1.57t

4.35t World
9.02t OECD

* countries chosen based on 2017 nominal GDP
@StatistaCharts Sources: International Energy Agency, International Monetary Fund

statista

Chart 10.5

If you are an American or a Canadian, please note that you are far and away the largest generator of emissions of any nationalities. The per capita emissions for both countries is 15 tons! This is more than double the per capita numbers of China and triple the numbers for the UK, Italy, France, and seven times higher than Brazil and India.

Business Sector Percentages of Total Emissions

Different segments of the economy will experience different degrees of difficulty in reducing their respective emissions totals. So some will require more time than others to reach their reduction goals. Here is how the total amount of greenhouse gas emissions breaks down:

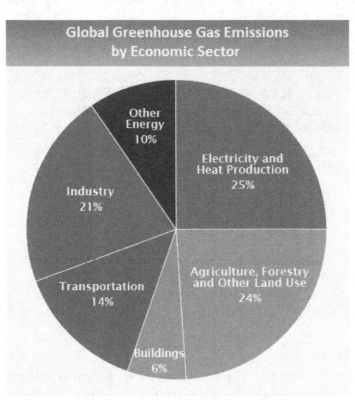

Chart 10.6 Source: US EPA[8]

When looking at this chart, there are two ways to strategize collective action. First it makes clear sense to tackle the biggest sectors of emissions first. That means starting with Agriculture/Forestry/Land Use and Electricity/Heat Production. Second, some sectors will require more time than others to get emissions under control. To radically reduce land use emissions might take longer than retrofitting buildings for lower emissions. So we also consider the low hanging fruit... where emissions can be reduced fairly quickly and easily.

The most straightforward position is to move on all sectors at once, keeping the overall goal in focus: lowering GHG emissions every year.

What must be remembered from Chapter 3 in Book 1 is the Resident C02 chart. C02 resides in the atmosphere for centuries so any emissions adds to the total. Said another way, if we had zero GHG emissions tomorrow, this Resident C02 would continue to warm the planet for decades.

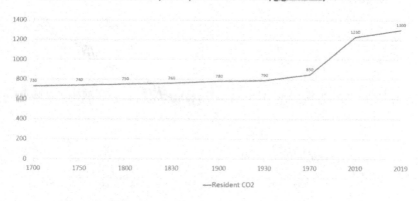

Resident CO2 in the Atmosphere (in billions of tons/gigatonnes) 1700 to 2019

—Resident CO2

Chart 10.7

The metric of annual C02 emissions reduction, along with Atmospheric Cleansing leads us to the last new metric.

Reduction of CO2 Parts Per Million (PPM)

As of this writing, the CO2 PPM in earth's atmosphere has reached the highest number in three million years. It is at 415 PPM. There's every reason to believe this number will increase at least throughout the end of 2019.

This is the critical issue relative to global warming and hence climate change. Even dramatically reduced annual emissions will add to this PPM number. Resident CO2 continues to accumulate as CO2 stays in the atmosphere for decades and even centuries. As stated in Book 1, the PPM issue can only be solved in the next decade by drawdown plus atmospheric cleansing. Drawdown reduces the amount of CO2 being emitted and added to the atmosphere, and atmospheric cleansing literally removes CO2 already resident in the atmosphere.

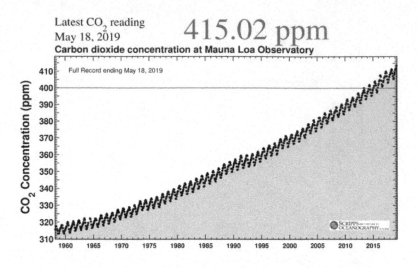

Chart 10.8

We must measure this number on an annual basis with the goal of reducing it to 350 or below by 2030

Annual PPM C02 Measurements

2020	416
2021	415
2022	414
2023	411
2024	405
2025	400
2026	390
2027	375
2028	355
2029	330
2030	310

The assumption in this chart is that we cannot deeply reverse the continuing increase in PPM numbers that has been going on for decades without Drawdown and Atmospheric Cleansing. We must remove more CO2 from the atmosphere than we are emitting into it by an average of 50 gigatons a year from 2020 to 2030. That means that the second half of the decade will have to overdeliver on the first half. So the 2024-2030 years will average almost 100 gigatons of CO2 reductions per year. This will a true test of our commitment to urgency in moving to a Finite Earth Economy.

Takeaways

1. New metrics are needed for the Finite Earth Economy.

2. As we will discuss in Chapter 11, the measurement of these metrics (and more) will increase in frequency and accuracy as terrestrial and satellite sensors are deployed, along with big data number crunching power and artificial intelligence enabled data analysis.

3. GDP as an indicator of the well-being of a nation is no longer a valid metric in the transition to a Finite Fuel Economy. GDP will be regarded as an old Growth Economy metric. CO2 emitted every year by humanity, moving back Earth Overshoot Day, and lowering the PPM of CO2 in the atmosphere will become replacement metrics to let us know how well we're doing.

4. Reducing total greenhouse gas emissions is essential. The countries with the greatest amounts of aggregated emissions must be responsible for the largest reductions.

5. The finiteness of Earth is being stressed by over consumption, over production and over population. We currently are consuming 1.7 earths worth of resources every year. We first overshot Earth's regenerative capacity in 1972

and over consumption has continued to increase every year since. We must cut consumption and waste ASAP to get back to one earth per year (or less).

6. Countries with both the highest percentage of energy derived from fossil fuels and the highest emissions must lead the way in reducing their use.

7. We strongly believe in Capitalism. We are not saying that humanity needs to move away from it. Capitalism is the correct economic model. That said, the metrics that were used to determine the health of a capitalistic economy will be rendered moot. In a Finite Earth Economy, capitalism will look quite different. The degree that Capitalism can adapt is the degree to which it will remain the dominant economic model.

Chapter Eleven

Essential Technologies

"Gentlemen, the President just told us that we need to do something that has never been done before, using technology that has yet to be invented, to do it in a short amount of time and that safety is a top priority. Let's get to work!" – attributed to James Webb, leader of NASA from 1961-1968, shortly after President Kennedy's 1961 speech about landing a man on the moon by the end of the decade.

Technology will play a critical role in moving to a Finite Earth Economy. Old, new, and yet to be invented technologies will all be needed if we want to succeed with the new metrics discussed in Chapter 10. The crucial need for urgency in targeting 2030 requires that we immediately fully deploy current relevant technologies. While we are doing that, we must motivate the government and private sector to invest in the development of new technologies.

In this chapter we discuss several categories of necessary technologies. While not exhaustive, we have a comprehensive list of the big technologies needed to help us "save ourselves from ourselves". In every case, we have links in Footnotes and/or in the Addendum should you want to dive deeper into any technology.

We start with a technology category that is essential to the success of the Finite Earth Economy: Energy.

Energy

The goal is to transition off fossil fuels ASAP. Almost every human alive has always lived in a culture and economy powered by fossil fuels. We must replace our perception of how life is lived with something new and different. We have to let go of legacy thinking across the board. We have no time to waste.

The author and the editor of these books strongly believe that humanity must move to a Finite Earth Economy as much as possible by 2030, or face the potential collapse of our civilization. To do that, we must deploy an **"all of the above"** strategy regarding energy. That is, every energy source that does not emit greenhouse gases must be developed to commercial scale and deployed as quickly as it safely can be.

Fossil Fuels

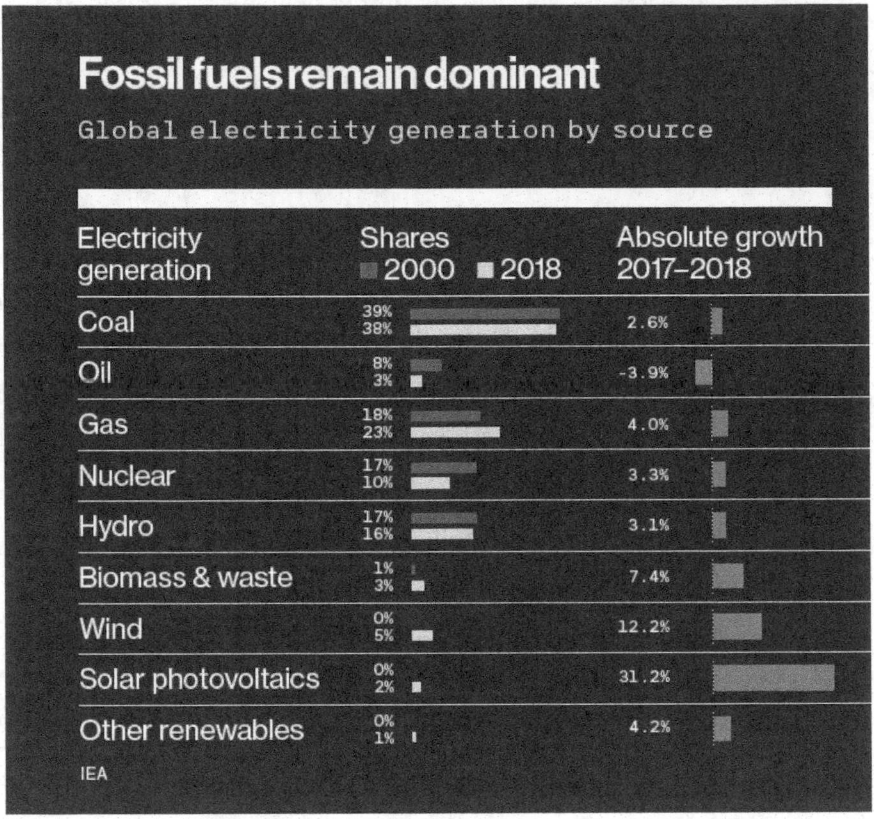

Fossil fuels remain dominant

Global electricity generation by source

Electricity generation	Shares ■ 2000 ■ 2018	Absolute growth 2017–2018
Coal	39% / 38%	2.6%
Oil	8% / 3%	-3.9%
Gas	18% / 23%	4.0%
Nuclear	17% / 10%	3.3%
Hydro	17% / 16%	3.1%
Biomass & waste	1% / 3%	7.4%
Wind	0% / 5%	12.2%
Solar photovoltaics	0% / 2%	31.2%
Other renewables	0% / 1%	4.2%

IEA

Chart 11.1 MIT Technology Review, May 2019

Fossil fuels will be in use to some degree for several decades. The near-term goal is to reduce fossil fuels in the total global energy mix to 30% by 2030, down from 77% today.[1]

NOTE: The chart above is specific to electricity generation. Fossil fuels are 64% of that mix. Transportation increases the fossil fuel share to 77%.

This means that the wealthier countries (which have the resources to do so) must decrease fossil fuels use below the 30% figure. This will allow poorer countries – who emit less greenhouse gas emissions a longer period of time to replace fossil fuels, and lessen the shock to their economies.

In addition to a wholesale transition to clean, renewable energy, capture of carbon emissions must be prioritized to drawdown as much as possible as soon as possible. The most in-depth and comprehensive analysis of Drawdown is the book by Paul Hawken: Project Drawdown.

In addition to coal and oil, natural gas is also a fossil fuel. It has primarily been portrayed (especially in the U.S.) as a relatively harmless alternative to oil and gas. This is not the case.

"The drilling and extraction of natural gas from wells and its transportation in pipelines results in the leakage of methane, the primary component of natural gas that is 34 times stronger than $CO2$ at trapping heat over a 100-year period and 86 times stronger over 20 years."[2]

Fracking has become a widely-used method to extract natural gas where traditional drilling is ineffective. Fracking is the process of injecting a slurry of water, chemicals and sand at extremely high pressure into fissures in the ground to release trapped natural gas. The byproduct is leaked methane, tainted ground water and, likely, earthquakes.

The argument is made that natural gas is not as bad for the environment as coal or oil. Being "less bad" does not equate to being good. Natural gas is not clean energy. It is less dirty energy than coal and oil.

"Everything is energy and that's all there is to it. This is not philosophy. This is physics." – Albert Einstein

"… if we want to know the secrets of the universe, we should focus on the non-physical aspects rather than physical ones, that will speed up the inventions." – Nikola Tesla

Simply stated, if everything is energy, why are we continuing to dig stuff out of the ground and burn it? Fossil fuels are physical energy extracted from the Earth. The fuels in this chapter derive energy from non-physical sources (i.e. Tesla's vision).

Wind and Solar

As this book is being written these two renewable energies are at the same relative cost as fossil fuels. The United States directly subsidizes the fossil fuel industry with $26 billion annually.[3]

The International Monetary Fund (IMF) found that direct and indirect subsidies for coal, oil and gas in the U.S. reached $649 billion in 2015. The study defines "subsidy" very broadly, as many economists do. It accounts for the "differences between actual consumer fuel prices and how much consumers would pay if prices fully reflected supply costs plus the taxes needed to reflect environmental costs" and other damage, including premature deaths from air pollution.[4]

The global subsidies by governments to the fossil fuel industry totals $5.3 trillion a year.[5] That number reflects the broader definition of "subsidy" used by the IMF and most economists.

Transferring all that money from fossil fuels to clean energy sources will significantly decrease the need for additional investments and enable a scaling up to quickly supply most of the world's needs. The legacy thinking of "cost too much" arguments are no longer relevant.

Solar and wind currently represent 7% of energy used globally[6] and 10% of U.S. energy.[7] These levels could increase by 500% by 2030, bringing the total to 35%. This quintupling in the next 11 years will require increasing investment and deployment. As with any technology, solar and wind will decrease in cost and increase in efficiency as they are ramped up.[8]

Both solar and wind technologies have increased dramatically in efficiency. Since1955 the percent of conversion of sunlight into energy with solar panels has increased from 2% to 15%. Wind energy efficiency has increased from 9% in 1980 to 40% in 2018.[9]

The cost to bring a megawatt of energy online is currently $1 million for Solar[10] and $1.75 million for Wind.[11]

These costs have dropped approximately 400% since 1980 and with investment and scaling efficiencies could credibly drop another 200% by 2030.

		2016	2017
INVESTMENT			
New investment (annual) in renewable power and fuels¹	billion USD	274	**279.8**
POWER			
Renewable power capacity (including hydro)	GW	2,017	**2,195**
Renewable power capacity (not including hydro)	GW	922	**1,081**
Hydropower capacity ²	GW	1,095	**1,114**
Bio-power capacity	GW	114	**122**
Bio-power generation (annual)	TWh	501	**555**
Geothermal power capacity	GW	12.1	**12.8**
Solar PV capacity³	GW	303	**402**
Concentrating solar thermal power (CSP) capacity	GW	4.8	**4.9**
Wind power capacity	GW	487	**539**
Ocean energy capacity	GW	0.5	**0.5**

Chart 11.2 REN21 – Renewables 2018 Global Status Report

Hydroelectric

Hydroelectric power currently supplies 18% of global energy usage.[12] In New Zealand, where hydroelectric supplies 60% of the power, the country has been able to completely go without fossil fuels for short periods of time. This has also occurred in Portugal.

Hydroelectric power is not likely to grow significantly because almost all of the rivers large enough to provide it have already been dammed and are already producing power. A second reason is that climate change is causing water-related problems. New dam construction faces return on investment and environmental hurdles.

Hydroelectric will remain the number one green energy source for the next few years, until Wind and Solar catch up. Wind, Solar and Hydro together currently supply 25% of the world's electricity. In moving to a Finite Earth Economy these three energy sources will need to be over 50% of all energy by 2030.

Geothermal

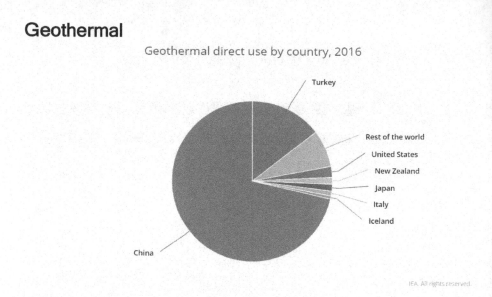

Geothermal direct use by country, 2016

Source: Renewables 2018

Chart 11.3

Geothermal currently comprises only 1% of global energy usage. It is used for heating and electricity in 70 countries. Iceland produces 27% of its electricity from geothermal. Due to its small population that translates to only 1% of the world's consumption of geothermal.[13]

It is possible to produce up to 8.3% of the total world electricity with geothermal resources, supplying 17% of the world population. Thirty nine countries (located mostly in Africa, Central/South America and the Pacific) could potentially produce 100% of their electricity using geothermal resources.[14]

The main advantage of geothermal heating and power generation systems is that they are available 24 hours per day, 365 days a year and are only shut down for maintenance. Geothermal power generation systems typically have capacity factors of 95%. Geothermal use will expand because it's clean with no emissions and it's relatively inexpensive: savings from direct use can be as much as 80% over fossil fuels. Geothermal's limitation is that it is available only in geographic areas where the earth's mantle is close to the surface.[15]

Ocean Energy

Ocean Energy refers to the renewable kinetic energy (i.e. motion) derived from waves and tides. In addition, surface water stores heat that can be harnessed to generate electricity. Researchers have been able to take advantage of the natural difference in pressure between saltwater and freshwater (known as "the salinity gradient") to generate electricity.[16]

The technology used to capture ocean energy falls within four major categories: wave energy converters, tidal energy converters, ocean thermal converters and salinity gradient technologies. These devices can capture energy and deliver electricity that is reliable, predictable and sustainable.

Experts project we could develop 748 gigawatts (GW) of ocean energy by 2050 worldwide. That level of capacity could save 5.2 billion tons of CO2 emissions.[17] We need to invest heavily in ocean energy to bring that milestone closer to 2030.

Nuclear

Since humanity as a whole is facing an existential crisis with climate change, we need to make some potentially unpopular decisions. As stated above, we believe that we must deploy every non-GHG emitting power source that we can. That includes nuclear power.

Many have a knee jerk reaction to the use of nuclear. They worry about accidents and the disposal of nuclear waste. The reality is that the number of people who have died or contracted cancer from all nuclear power plants pales in comparison to deaths attributed to the extraction and burning of fossil fuels.

Every day 10,000 people die from issues directly related to the use of fossil fuels, primarily air pollution (also mining and drilling accidents, water pollution, transportation and pipeline issues). In a single year more people die from the burning of carbon fuels than have died from nuclear energy throughout its entire history. So, the idea that we should not embrace nuclear because it is dangerous, must be refuted by the facts.[18]

We cannot allow legacy fears and emotions concerning nuclear power to shutter reactors currently supplying clean power, or to prevent us from developing and deploying new technology reactors that will be smaller, safer and more distributed. Reaching 70% clean energy by 2030 is critical to humanity's survival; and we can't get there without taking advantage of nuclear power.

The Chernobyl explosion caused two immediate deaths and 29 deaths from acute radiation sickness in the course of the next three months. 50 people died of acute radiation syndrome, and 4,000 may ultimately die of radiation-related causes.[19]

The Fukushima Daiichi nuclear disaster caused no deaths due to acute radiation exposure. Studies by the World Health Organization and Tokyo University have shown that no discernible increase in the rate of cancer deaths is expected. Predicted future cancer deaths due to accumulated radiation exposures in the population living near Fukushima have ranged in the academic literature from none to hundreds.[20]

Numerous studies around the world have concluded that there are no health risks to people who live near nuclear reactors. Compare this to the World Health Organization findings that urban outdoor air pollution, from the burning of fossil fuels, is estimated to cause more than **three million deaths worldwide per year** and indoor air pollution from biomass and fossil fuel burning is estimated to cause approximately 4.3 million premature deaths.[21]

Add in all the deaths from oil drilling and refinery accidents, and nuclear-related health concerns fade. Every time a nuclear reactor is decommissioned, there is an immediate rise in CO_2 emissions. The energy has to be replaced by something, and the replacement is typically coal.

In the writing of this book, we were lucky enough to get a personal introduction to Herschel Specter. Here are some highlights from his bio: Herschel Specter, President of Micro-Utilities, Inc., holds a BS in Applied Mathematics from the Polytechnic Institute of Brooklyn and an MS from MIT in Nuclear Engineering. He was a member of the Atomic Energy Commission (now the Nuclear Regulatory Commission) where he was the Licensing Project Manager for the original licensing of the Indian Point 3 nuclear plant. He served at diplomat rank for five years at the International Atomic Energy Agency in Vienna, Austria where he led an international effort writing design safety standards for nuclear power plants.

Herschel knows whereof he speaks. Here's what he wrote in his 2019 report "Become a Nuclear Safety Expert":

"The bottom line of all this is: In US designs, and those of many other countries, severe nuclear power plant accidents are rare and extremely unlikely to cause any near term off-site radiation fatalities or radiation sicknesses. Long term effects, if any, would be too small to be detected. We also know today that the risk of contaminating land areas from a nuclear accident is far less than thought before because only very small amounts of cesium would be released and because natural

"weathering" effects rapidly reduce cesium dose rates. Extreme claims by some about nuclear accident consequences are not supported by advanced analyses or by actual nuclear accidents. Nuclear power plants operating today do not represent a significant threat to society. Future nuclear plant designs will do even better as many new designs will avoid reactor meltdowns altogether... Carbon-free electricity from nuclear power plants is essential in dealing with climate change."[22]

That isn't to say that there are no issues with nuclear. Currently running reactors must be regularly inspected and safety regulations adhered to. Any reactors vulnerable to sea level rise or other issues may be decommissioned.

Another nuclear success story devoid of accidents, sickness or death is the U.S. Navy. Under the stubborn and relentless leadership of Admiral Hyman Rickover (a safety fanatic) there has never been any problem of any sort with the nuclear powered vessels:

"U.S. Nuclear Powered Warships (NPWs) have safely operated for more than 50 years without experiencing any reactor accident or any release of radioactivity that hurt human health or had an adverse effect on marine life. Naval reactors have an outstanding record of over 134 million miles safely steamed on nuclear power, and they have amassed over 5,700 reactor-years of safe operation. Currently, the U.S. has 83 nuclear-powered ships: 72 submarines, 10 aircraft carriers and one research

vessel. These NPWs make up about forty percent of major U.S. naval combatants, and they visit over 150 ports in over 50 countries, including approximately 70 ports in the U.S.'[23]

Yes, the nuclear waste issue must be solved. It's a Not In My Back Yard (NIMBY) issue, but we must come together within the climate crisis reality we are in, to solve the problem. No country can claim to have a comprehensive solution for dealing with its toxic waste. Environmental group Greenpeace estimates that there's a global stockpile of about 250,000 tons of toxic spent fuel spread across 14 countries, based on data from the International Atomic Energy Agency. Of that, 22,000 cubic meters (roughly equivalent to a three-meter tall building covering an area the size of a soccer pitch) is hazardous, according to the IAEA. A 2015 report by GE-Hitachi put the cost of disposing nuclear waste at over $100 billion.[24]

One potential solution to the waste issue is to bombard the waste with short bursts of intense laser light thus changing the atomic structure of the nuclear waste. Gerard Mourou – one of the three winners of the 2018 Nobel Prize for Physics – claims that the lifespan of radioactive waste could potentially be cut to minutes from thousands of years.[25]

Existing nuclear technology includes Generation 1 and 2 nuclear reactors. These were all built in the 20th century electric grid paradigm. Next generation nuclear reactors are currently in development. Reactor suppliers in North America, Japan, Europe, Russia, China and elsewhere have a dozen new nuclear reactor designs at advanced stages of planning or under construction, while others are at a research and development stage. Fourth-generation reactors are at the R&D or concept stage.

Third-generation reactors have:

- A more standardized design for each type to expedite licensing, reduce capital cost and reduce construction time.

- A simpler and more rugged design, making them easier to operate and less vulnerable to operational upsets.

- Higher availability and longer operating life – typically 60 years.

- Further reduced possibility of core melt accidents rated at calculated core damage frequency (CDF) of 1×10^{-5}.

- Substantial grace period, so that following shutdown the plant requires no active intervention for (typically) 72 hours.

- Stronger reinforcement against aircraft impact than earlier designs, to resist radiological release.

- Higher burn-up rates to use fuel more fully and efficiently, and reduce the amount of waste.[26]

There are a number of Generation IV nuclear technologies currently in development. These are overseen and coordinated by the Generation IV International Forum. Currently they include:

- Very-High-Temperature Reactor (VHTR)

- Molten Salt Reactor (MSR)

- Supercritical-Water-Cooled Reactor (SCWR)

- Gas-Cooled Fast Reactor (GFR)[27]

All of these types of Gen IV are expected to be deployed by 2030. This work should be accelerated so that one or more of these technologies can begin deployment by 2025.

Better yet, the advanced nuclear reactors that will replace our existing reactors use up much more of the fuel than present-day reactors leaving less waste. Small modular reactors can better tailor energy use to demand. Molten salt reactors obtain up to ten times the energy from the same amount of uranium because the fission products and reaction poisons are removed as it goes. Fast reactors obtain at least ten times the energy from existing nuclear fuel by burning everything, not just U-235.

Using existing uranium from U-mine sites, as well as burning existing spent fuel in fast reactors in the near-future, provides sufficient uranium fuel to produce 10 trillion kW/year for thousands of years, making it presently sustainable by any measure.

Using uranium extracted from seawater, instead of mining uranium ore, makes nuclear truly renewable as well as sustainable. The amount of uranium in seawater is only 3.3 micrograms/liter (parts per billion), but that totals 4.5 billion tons of uranium in the billion cubic kilometers of seawater in the ocean.[28]

Nuclear will be an integral technology on our journey to a Finite Earth Economy by 2030. The arguments for nuclear are rational and based upon hard data and critical need. The arguments against it are based on emotion and misperceptions.

Space-Based Solar Power

Artist's concept of the transmission of space-based solar power. Image via JAXA.

Solar arrays, no matter how efficient, only generate power when the sun is shining. To circumvent that limitation, it is possible to place the arrays where the sun shines all the time, and where there are no atmospheric interferences (e.g. clouds) that reduce the intensity of the sunlight. Outside of the earth's atmosphere, orbits exist where there's no nighttime. So we can harness solar power that flows continuously.

The potential benefits are enormous. Space-based Solar Power (SBSP) is an enabling technology that could leapfrog the electric-power transmission grid on Earth, and have the same effect that the cellular phone system had on communications. Ground stations will be receivers for the radio-frequency power and will convert it into electrical energy. Some of these receivers will feed into existing electric utility grids. Others may be isolated micropower networks at remote off-grid locations or even individual farms or factories, all receiving a continuous infusion of power from the sky 24 hours a day.[29]

It will take vision, massive investment, research and development, time and energy to implement SBSP. Solar energy captured will be beamed down from space to receiving stations on earth. Microwave power transmission is the typical choice in SBSP designs due to general efficiency, but utilizing laser power beaming is another option due to lower weight and cost. There is a potential for misuse because someone could turn either choice into a space weapon. Safety protocols can easily deter this unlikely threat. A simple handshake between antenna and rectenna would disable the transmission if it veers off course.[30]

Energy conversion rates will be determined by the efficiency of the solar arrays (currently at 15%). By the time a SBSP can be deployed, it's safe to assume efficiency rates will be approximately 30%. On the ground microwave efficiency conversion rates are difficult to forecast at this point, but scientists don't consider them to be a major hurdle.[31]

Two keystone constructs are required for space solar to work. First are solid-state power amplifiers that efficiently convert electricity from collected sunlight into radio-frequency waves, and second are receivers on the ground that re-convert the RF waves back into electricity.

"If you sit outside on a sunny afternoon for 15 minutes, you don't get burned. Our radios, TVs and cell phones aren't cooking us, and those are all at the same frequencies as what's being proposed. There are already safety limits (on microwave transmissions) set by the IEEE (Institute of Electrical and Electronics Engineers), so you design a system to make sure the power is spread over a large area. It won't accidentally turn into a death ray." – Paul Jaffe, an engineer at the Naval Research Laboratory in Washington, D.C.[32]

To get the best cost-to-weight ratios, efficiencies of scale, and experience comparable electrical generation capacity of an average nuclear power plant (1 to 2 gigawatts), any solar collection array in space would need to be roughly a kilometer in diameter.

Collection receivers on the ground would need to be accordingly large – for a space-based solar plant to generate around one gigawatt of energy, a one-kilometer (.62 mile) solar collector would beam energy to a 3.5-wide kilometer (2 mile) receiver on the ground. That would require an area of around 900 acres. Compare that with the Solar Star solar panel plant in California, currently the United States' largest solar utility, which occupies 3,200 acres and generates .58 gigawatts.[33]

Radio-frequency power transmission does have one significant drawback: the "safe" wavelengths that also won't get refracted by something as simple as rain are already overcrowded, clogged up with regular radio transmissions, as well as military, industrial and satellite use.

Critics of space solar (prominent among them Tesla's Elon Musk), say economy-scale efficiencies just can't be achieved because of all the converting and reconverting of the power that is required. Paul Jaffe and other scientists feel that the crowded frequency and conversion issues can be overcome.[34]

SBSP falls into the category of a big solution, a moon shot and a real option to be considered going forward. It has within it the opportunity to be largely a single solution in several decades. The issue of course is the hefty cost of deployment, north of $1 trillion to initially implement. This is within the context of tens of trillions of dollars in the energy sector alone to globally convert to a Finite Earth Economy. That amount will still be significantly less than the hundreds of trillions that will be lost in the next 50 years if we do not convert to a Finite Earth Economy.

SBSP is a source of energy that should be fully researched and tested so that by 2025 we can move to deployment.

Energy Storage Technology

In the last 125 years, battery technology has advanced exponentially. Think of the big early transistor radios compared to today's smart phones. Both powered by different forms of batteries. In early stage development of the lithium ion battery, the initial problem was increasing performance while decreasing size/weight and lowering cost. One reason that electric cars cost more than their internal combustion counterparts is the price of the batteries.

Elon Musk leads the world in electric automobiles with his Tesla brand. He has single handedly altered the automotive landscape. Tesla cars cost more due to the cost of the batteries. In addition, scaling up production was limited by the available supply of batteries. So Musk built the Gigafactory to have a plentiful supply of batteries for his automotive manufacturing needs. The chart below shows how big the first Tesla Power Gigafactory is. There are plans to build several more around the world.

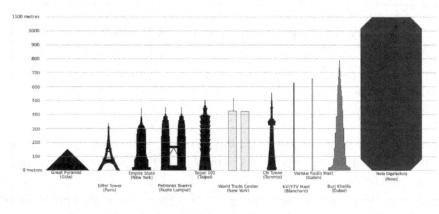

Chart 11.5

Tesla Power is the first of what will be numerous other companies in this new storage market. Here is a picture of the first Gigafactory outside Reno, NV. It is fully powered by wind and solar power. You can see the solar panels on the rooftop and the wind turbine farm in the upper right.

Chart 11.6

This is a huge end to end energy solution! Think about how solar and wind are variable sources of energy. On sunny or windy days they create power, often in excess of current need, so they provide it back to the grid. When solar and wind are not producing enough power, the grid supplies the extra power needed, meaning the gaps are filled by fossil fuel sources.

With scaled up battery storage, when there is excess wind or solar power, it is stored to be used on sunless or windless days. So ever more efficient, ever cheaper and larger battery storage systems solve the variable issue of solar and wind energy.

To show this potential, Musk made a big, risky move in Australia. He bailed out the Province of South Australia by building the world's largest battery to provide the province with Frequency Control Ancillary Services (FCAS). To quote from an Australian Broadcasting Company article:

"The Hornsdale Power Reserve is registered to provide what is known in the power markets as Frequency Control Ancillary Services (FCAS). FCAS requires providers to keep a little bit of power in reserve — which the market operator can use to help correct the supply/demand balance in response to minor changes in load or generation.

Some FCAS services are reserved for use in a major event like a major power station fire, a transmission line tripping or a big industrial load switching off. Until Tesla's big battery switched on, FCAS services in Australia had only ever been provided by traditional coal, gas, diesel and hydro generators."[85]

In addition to replacing fossil fuels, the battery has resulted in tens of millions of dollars of savings in its first year.

In a Finite Earth Economy, battery technology will not only be utilized for powering our devices, it will fill in the energy gaps of variable energies such as Wind and Solar. As proven in South Australia, it can also be a back-up energy source to the electricity grid.

Future Sources of Energy

Nuclear Fusion

This might well be a new, major source of carbon-free energy in the coming decades. Currently 35 nations have joined forces in ITER, a group whose commitment is to bring Fusion to the market by the mid-2030s. As this is the cutting-edge effort in this energy category, we quote from their website:

"ITER ("The Way" in Latin) is one of the most ambitious energy projects in the world today. In southern France, 35 nations are collaborating to build the world's largest tokamak, a magnetic fusion device that has been designed to prove the feasibility of fusion as a large-scale and carbon-free source of energy based on the same principle that powers our Sun and stars.

The experimental campaign that will be carried out at ITER is crucial to advancing fusion science and preparing the way for the fusion power plants of tomorrow.

*ITER will be the first fusion device to produce **net energy (produces more energy than what was required to generate it)**. ITER will be the first fusion device to maintain fusion for long periods of time. And ITER will be the first fusion device to test the integrated technologies, materials, and physics regimes necessary for the commercial production of fusion-based electricity.*

*Thousands of engineers and scientists have contributed to the design of ITER since the idea for an international joint experiment in fusion was first launched in 1985. The ITER Members – **China, the European Union, India, Japan, Korea, Russia and the United States** – are now engaged in a 35-year collaboration to build and operate the ITER experimental device, and together bring fusion to the point where a demonstration fusion reactor can be designed."*[37]

This form of energy will not help us get to a Finite Earth Economy by 2030, but it has within it the promise of being a dominant energy source for the second half of this century.

In summary, we strongly support the "all of the above" strategy for clean energy. All forms of clean energy will be needed to reach the goal of reducing energy from fossil fuels to only 30% of all energy required by 2030. The "green movement" has been counterproductive by engaging in intramural fights about what are acceptable sources of energy. These fights have not assisted rapid deployment of all clean energy sources. It's time to pull together and deploy every clean source of energy we can in order to reduce the burning of fossil fuels ASAP. 2030 is our goal for a 30% limit of fossil fuels. There's no time for squabbling.

Agricultural Technologies

The agriculture business sector is the largest producer of greenhouse gas emissions and therefore a top priority to change how and where food is grown, and what technologies might help in the lowering of emissions.

In the 20th century we moved from a multi-crop agricultural paradigm to an industrial model. Most food today is not locally grown, it must be transported (sometimes thousands of miles) to reach our kitchen tables.

We must redesign how and where food is grown, and to grow it using only renewable energy. Here are the technologies that will help in this critical area as we move to a Finite Earth Economy.

Regenerative Permaculture

"Farming the land as if nature doesn't matter has been the model for much of the Western world's food production system for at least the past 75 years. The results haven't been pretty: depleted soil, chemically fouled waters, true family farms all but eliminated, a worsening of public health and more. But an approach that combines innovation and tradition has emerged, one that could transform the way we grow food. It's called agroecology, and it places ecological science at the center of agriculture. It's a scrappy movement that's taking off globally. Much of the world is waking up to the costs of the industrial approach that defines most of American agriculture, with its addiction to chemicals and monoculture. A new reckoning known as true cost accounting is putting dollar figures on industrial agriculture's contribution to soil erosion, climate change and public health."[38]

Lab-based Meat

The production of meat, particularly beef, comes at a high cost. It is the least efficient way to deliver nutrients to humans. It generates massive amounts of greenhouse gases in its production and transportation. Other undesirable factors of meat production include that animals must be slaughtered, and the antibiotics and growth hormones given to them at factory farms are then consumed by humans.

Lab-based meat is artificial, clean meat that is grown in laboratories. As a futurist, I remember talking about this 10 years ago and in that time, it has gone from early stage experimentation to a sophisticated science that is ready to be scaled up for markets.

During the writing of this book, it was announced that Burger King is introducing a burger made largely from pea protein and testing it in its St. Louis area stores. While not lab-grown meat, it is a huge step forward in the effort to both cut carbon emissions from agriculture and stop the slaughtering of livestock.

If current meat production processes were replaced with lab-grown meat, it is estimated that 40% less energy would be required, 80% less land would be needed, and 80 to 90% less water would be used. These are three significant reasons for investing heavily in the research and development of lab-based meat.

Transportation

Automotive

100 years ago, society moved from horse and buggy to internal combustion engine. Now we are moving from internal combustion to electric and perhaps hydrogen powered vehicles. The technology has been proven. The cost must be reduced to make electric and hydrogen vehicles affordable to the masses. That will happen with scale. Sales of electric vehicles (EVs) are scaling up rapidly.

As of 2017 there were three million EVs on the roads worldwide.[40] This was a 50% expansion over 2016 – a huge increase, but in the larger scheme of things it is still very small. Currently there are approximately one billion cars on the road globally.[41] Of that the US has approximately 250 million cars, about one million of which are EVs. We are in the infancy of a technology that must rapidly scale up for us to reach a Finite Earth Economy by 2030.

How can this technology be scaled up in time?

First, we can anticipate a continuation of the rapid year over year expansion.

Second, the growing immediacy of climate change is creating a much greater awareness of the need to transition from fossil fuels.

Third, if governments feel pressure from their constituents, they can subsidize purchases. This has been successful in the past. Now with a growing awareness of the tens of trillions of dollars of economic loss looming due to climate change, the ROI of subsidies is compelling. The global goal should be to have 400 million EVs plus 200 million hybrids on the roads by 2030.

Autonomous Vehicles

Autonomous vehicles are cars, buses and trucks that operate without a driver. They have begun to navigate highways around the world and are poised for explosive growth.

Again, we are confronted with legacy thinking. The technology is viable, but there will be a lag time before the public, and many legislators, are ready to accept cars without drivers on the open roads. Here are some statistics that may help. 97% of all driving fatalities globally are due to "driver error".[42] Autonomous autos will be much safer as the most dangerous, most easily distracted component of the driving experience is replaced.

The other major statistic is that in the U.S. the average car is used only 5% of the time.[43] People use them to drive to work, where the cars sit all day in a parking lot, and then they drive home where the cars sit in a driveway or garage depreciating. Does it make sense to have one of our largest personal investments sitting idle 95% of the time? Of course not! However, we have all grown up in a car culture, so it is an automatic decision to have one.

Now, envision a reality where there are autonomous cars constantly on the streets, roads and highways 20 to 22 hours a day (autonomous EVs will drive themselves to a charging station a couple of times a day). They are always available to people who need a lift. This means that by 2030 there could be half as many cars on the road as today. People would still be able to get where they need to be when they need to get there… and they'll spend a lot less to do so.

This brings us to a critical point about infrastructure. During the last 10 to 15 years there has been considerable discourse around the need to have "green infrastructure" or "green transportation initiatives". This futurist has been on several panels and has written extensively about it. Once I looked deeply into autonomous autos a few years back the "aha" moment hit me! The thinking on this was backwards. "What new infrastructure do we need?" Why change the infrastructure when we're already changing the vehicles? We don't need to spend hundreds of billions on light rail or some other costly endeavor. We have the green infrastructure in place: the Interstate Highway system! With half as many cars on the road, traffic will be a non-issue.

This will allow governments at all levels to save money on unnecessary infrastructure projects and instead invest in helping citizens transition from a car culture to a shared vehicle culture.

Air Travel

The good news about air travel is that it has roughly an 81% occupancy rate, making it the most efficient in terms of passenger load. The bad news is that air travel accounts for approximately 12% of all GHG emissions. Personally, given that I fly a lot, it is the single biggest contributor to my carbon footprint.

In the past couple of years there has been a nascent development of electric batteries for air travel. At the time of the writing of this book, there are small commercial enterprises flying up to 150 miles using batteries, which get replaced upon each landing. By the early 2020s, a sizeable percentage of flights under 500 miles could be powered by removable batteries. 75% of all business plane trips are less than 250 miles in the U.S.

The power of governments to fund and drive these transportation sector changes in the effort to convert to a Finite Earth Economy can result in a 50 to 75% cut in GHG emissions by the aviation industry by 2030.

Technological Environmentalism / Fourth Wave of Environmentalism

"As the business world enters the Fourth Industrial Revolution, the environmental community has been heading into our own complementary Fourth Wave of environmentalism. The land-conservation movement of the early 20th Century was the First Wave, and the second, which began in the 1960s, used the force of law to reduce industrial pollution. The Third Wave emphasized the power of economics — problem-solving, market-based approaches, and corporate/NGO partnerships.

Now the Fourth Wave is seeing technological innovations giving people new powers to scale up solutions by supercharging all of the approaches that came before." — Fred Krupp, President, Environmental Defense Fund

Emerging and powerful new technologies are rapidly being deployed around the world to better measure and understand what is happening now, and forecast what we can expect on the planet, in the oceans and in the atmosphere. These technologies will truly transform business and society as they enable much better analysis of how Earth is responding to the way humanity lives on it.

Here are some of the technologies:

- Additive Manufacturing
- Automation Technology
- Big Data and Data Analytics
- Blockchain
- Internet of Things
- Satellite Imaging and Communications
- Sensors
- Sharing Technologies

NOTE: Further information on each of these technologies is available in the Addendum.

These technologies will reframe climate conversations as we will be able to bring real time global data into daily evaluation, and analyze these truly massive amounts of data to make sense of them, run simulations against them and make informed decisions concerning the ultra-complex, hyper-connected system that is our climate.

These technologies will help immensely in the fundamental shift from Growth Economies to Finite Earth Economies. If it can be measured, it can be monitored and analyzed. We will now be able to measure emissions, determine their points of origin, and accurately place blame when necessary – on a daily basis!

Just as we have been able to reduce crime in major cities through the use of vast networks of CCTV [closed circuit TV] cameras, satellites are beginning to watch Earth from above to monitor GHG emissions by country, and even with more pinpoint accuracy. Recent increases in CFC-11 emissions have successfully been traced to foam manufacturing plants in the Chinese provinces of Shandong and Hebei. Those factories have been shut down and CFC-11 levels in the atmosphere are dropping.[45] New satellites are creating a global CCTV network that will act as a watchdog on all governments and corporations. Those countries and businesses will know their emissions are being tracked daily.[46]

Drawdown, Carbon Capture, Sequestration, Carbon Offsets and Direct Air Capture Technologies

Drawdown

This is the strategic, data driven effort to reduce existing and future GHG emissions by reengineering how we do things and what we can use as intelligent replacements to the current carbon intensive reality of today. It can be used as an umbrella term for any technology or process that reduces carbon emissions (whether at the source or by removing CO2 from the atmosphere).

Carbon Capture

This is the process of capturing carbon that is a byproduct of fossil fuel production. It is primarily used in the processing of "clean coal".[48] The current limitation to the growth of its use is that a C02 aftermarket has not fully developed. In addition to reuse, the carbon that is captured can also be sequestered. This technology must be further developed to increase efficiency and to lower cost. Then it must be deployed to every source of fossil fuel production. ASAP.

Sequestration

This is the back end of carbon capture. It takes C02 and sequesters it in the earth, deep underground or elsewhere in a fashion that prevents the CO_2 from escaping back into the atmosphere. Carbon sequestration happens naturally. Most of the CO_2 emitted is absorbed into the oceans or is used by trees and other plants in their photosynthesis process. The issue today is that these natural carbon sinks are overwhelmed by the magnitude of our CO_2 emissions. Planting trees is a great idea and helpful, but we must do more to capture and sequester CO_2.

The technology, infrastructure and expertise that the oil and gas industry possess and use for extraction can also be used for injection. This industry knows that their output will be winding down in the next several decades, so why not provide them with incentives to help reverse the damages they have profited from? Occidental specializes in using CO_2 injections to boost production from oil wells. So, it's a simple first step for the oil and gas industry to capture CO_2 emitted during the drilling/fracking processes and then inject that CO_2 deep into the earth where it will mix with water and minerals to create carbonite – a solidified form of CO_2. The result is sequestered CO_2 that will not seep back into the atmosphere. We suggest that the fossil fuel industry be required to include this additional step into their cost of doing business.[48]

Carbon Offsets

As will be laid out on Book 3, in moving to a Finite Earth Economy, tax codes and government policies will be reoriented to both reducing carbon and offsetting the CO_2 generated. The idea is that if I do not have a carbon neutral footprint, then I can buy or invest in a carbon capture plan to offset my carbon emissions. Paying for the planting of trees is a simple example. A tree can consume up to 48 pounds of C02 per year.[49] To improve on that, artificial trees are being produced and sold that can absorb more than 1000 times that.[50]

The other option is to invest/buy into one of the above carbon technologies as an offset. To put the magnitude of the problem into perspective, the average U.S. citizen is responsible for 16 tons of C02 per year.[51] So all kinds of new CO2 absorbing technologies must be created. Thirteen years ago I wrote about titanium dioxide, a whitening agent that, when painted onto concrete, actually absorbs C02 from the air up to eight feet away from the paint.

All kinds of offsets could be created whereby people, companies and governments can offset their GHG emissions. For these emissions reduction projects to generate carbon offsets, they must meet a series of requirements including: internal monitoring, external verification, and a system of accountability (should the project fail to accomplish its climate impact goal).

Direct Air Capture

Direct air capture scrubs CO2 directly from the ambient air. Removing CO2 to reduce the amount resident in the atmosphere is a major goal. It is being done on a small scale in tests around the world. The major hurdle is deploying it at a scale that will make a difference.[52] The technology remains costly and energy-intensive. It is often difficult to pin down costs for new technologies, but one recent study estimates that it would cost about $94-$232 per metric ton. Previous estimates were higher.[53]

Direct air capture also requires substantial heat and power inputs – scrubbing one gigaton of carbon dioxide from the air would require about seven percent of all projected U.S. energy production in 2050. The technology would also need to be powered by low or zero carbon energy sources to result in net carbon removal.

Multiple companies have already developed working direct air capture systems, despite the near absence of public research and development spending on the technology. The bottom line is that direct air capture is still a new technology, and these systems are just the first of their kind.[54]

Atmospheric Cleansing

This is a future technology to reduce C02 in higher altitudes of the atmosphere. There has been research done on how the upper atmosphere cleanses itself of pollutants and that it can be a balanced process. Now however, the increase from 730 gigatons of resident C02 in the atmosphere pre-industrial revolution to the current 1300 gigatons means that if we want to realize a Finite Earth Economy by 2030, we must counteract the increasing amount of annual emissions by actively cleansing the higher atmosphere.

It is worth noting that around the time that resident $CO2$ became 830 gigatons in the early 1980s is when the planet began its warming. This was also around the time that we exceeded 350 PPM. This technology must be deployed in the short term of 10-20 years to accelerate retuning to $CO2$ stasis. This is one of the fastest ways to reduce the PPM of C02 in the atmosphere as it captures the resident C02.

I wrote about this in 2012:

"The transformative technology that might well come along within the next few years would go beyond the tree model and move to the "vacuum cleaner" model: actively sucking C02 out of the atmosphere. This is an environmentalist's holy grail technology. It will be developed during the Shift Age. The question is when it can be scaled up to a global level of impact so that this technology can actually overcompensate for the C02 humanity is putting into the atmosphere. This could well happen by the 2020s, and if so might be one of the more influential technologies of the Earth Century, of the Anthropocene Era." – Entering the Shift Age, Houle 2012

This is a call to Silicon Valley and other pockets of technological disruption to create a technology that can cleanse the atmosphere above 10,000 feet. The wealth created from developing a technology that makes a significant contribution to saving humanity will be gargantuan. This is needed ASAP to lower the PPM of C02 in the atmosphere. Once we do, this technology may no longer be needed if we have lowered our annual emission levels to single digit gigatons from its current 37 per year.

Seawater Capture

Seawater capture is akin to direct air capture, except CO_2 is extracted from seawater instead of air. By reducing CO_2 concentration in the ocean, the water then draws in more carbon from the air to regain balance. Seawater has a more concentrated solution of CO_2 than the ambient air, which means less work is required to separate it out than in direct air capture. But seawater is also considerably heavier than air, which means more work to move it through the system. Seawater capture will also have to grapple with the added complexities of technology deployment in harsh maritime environments.

The U.S. Navy has already developed a prototype seawater capture device. Because CO_2 can be converted to fuel and avoid having to stop to refuel. Of course, if the captured carbon is converted to fuel and combusted, it returns to the atmosphere, but future applications of this kind of technology could provide long-term storage for captured carbon.[55]

Enhanced Weathering

Some minerals naturally react with CO_2, turning carbon from a gas into a solid. The process is commonly referred to as "weathering," and it typically happens very slowly – on a geological timescale. The weathering process can be accelerated by dispersing magnesium iron silicate powder on a land surface. It then draws CO_2 out of the atmosphere and sequester it in newly-formed rock mineral, magnesium carbonate. Scientists have shown that enhanced weathering works, but there is more to be done to determine if it can cost-effectively scale, and if there may be unintended consequences.[56]

High Risk Technologies

Geoengineering

Geoengineering is the creation of massive engineering projects to significantly alter Earth's atmosphere to offset global warming. For example, deploying sun shields protecting large swaths of the earth from the heat of the sun to reduce terrestrial temperatures. It is impossible to assess the ratio of risk to rewards with these types of projects.

Earth's biosphere is mind-bogglingly complex – a gargantuan, integrated system. There's no way to test in a laboratory what unintentional consequences might occur from geoengineering projects.

If we find over the next several years that we have moved past a major tipping point… that we are hurtling toward a runaway hothouse effect, geoengineering may be worth the risk. Until then, we will shelve them and focus on current technologies and moonshots like Space-based Solar Energy and Atmospheric Cleansing.

Takeaways

1. Humanity must reduce fossil fuel usage from the current 77% of all energy to 30% by 2030. To do so we must utilize ALL forms of clean energy.

2. All governmental fossil fuel subsidies must be transferred to subsidizing only clean energy sources. Clean energy can no longer be claimed to be more expensive than dirty.

3. "Moon shot" technologies such as SBSP and Atmospheric Cleansing must immediately go into funded research and development so that both can be deployed by 2025.

4. Energy storage technology is critical to the growth and use of variable energies such as Wind and Solar.

5. Decreasing the cost of batteries is key to transitioning all land vehicles to electric. This will also allow electric air travel to become common for short range air travel.

6. Given the fact that C02 resides in the atmosphere for decades and centuries, there must be a full, well-funded effort to rapidly scale up carbon capture at all levels of the atmosphere.

7. Carbon sequestration can be accelerated by deploying existing oil and gas extraction technologies to inject captured CO2 into dormant wells, and by requiring the oil and gas companies to absorb the cost.

8. A carbon after-market needs to be established so captured carbon can become a revenue stream. Once the amount of carbon removed from the atmosphere truly scales up in tonnage there needs to be a price paid for the sequestered carbon. This will provide a future revenue stream for the fossil fuel industry to replace revenue lost from oil and gas production. This is a true form of 21st century retrofitting.

Chapter Twelve

Timelines and Attainable Milestones

"Time is the scarcest resource and unless it is managed, nothing else can be managed." – Peter Drucker

This is the toughest chapter of this book to write. There are so many variables involved, it makes the setting of timelines and truly attainable milestones an extremely difficult undertaking.

First let's take a look at the reality of where the world is relative to climate change.

1. Climate change denial is largely in the rear-view mirror globally, except for the President of the United States, and many of his followers. Denial, so strong even five years ago, is being tossed into the dustbin of history due to the reality set forth in Book 1: that climate change is here and now, an experiential reality. In addition, the number and size of climate crisis demonstrations worldwide has garnered widespread attention including in the media. This is putting pressure on politicians to address climate change. Ballot initiatives are starting to work as well.

2. The energy, transportation and financial sectors are all in fundamental transition from the old to the new, from fossil fuels to alternative, clean sources of energy. This transition has shown progress in recent years, but the fundamental truth is that GHG emissions keep going up. Loosely, a lot of talk but no walk.

3. There is not deep widespread awareness of cause and effect relative to individual and collective actions. For example, when eating a quarter pound cheeseburger, the vast majority of people are not aware that six pounds of CO_2 were emitted in the production of that cheeseburger. I think that less than one percent of Americans know that, on a per capita basis, each of us is responsible for 16 tons of GHG emissions per year.

4. When does the will to collectively face our common enemy – climate change – trigger a global and national mobilization that is up to the task? Americans most often refer to the unprecedented ramp up at the beginning of WWII as an example of how quickly we can come together and defeat a common enemy. Can the US Democracy move fast enough?

Will authoritarian governments around the world join forces with the rest of us to defeat this common enemy?

To comment on the above four realities:

Number 1 must continue at all levels of demonstration, petitions, voting, lobbying. The media must treat this topic with the level of importance and amount of coverage it deserves. I am constantly frustrated and disappointed with the cliched and superficial media coverage of climate change. (Open with a shot of belching smokestacks, cut to ice melting, have a concerned reporter mention starving polar bears.) This is not a "story" but humanity's collective future at ominous risk. The people are ahead of the leaders, which has been the case on this for decades. How quickly can "the people" move the levers of power?

The pace of change relative to Number 2 must accelerate. Since much of this is economic, the Finite Earth Economy must enter the collective consciousness as the best, healthiest, most financially rewarding choice we can make. Until we achieve three consecutive years of declining emissions, we are simply talking desire and not walking reality.

Number 3 is the reason we feel so strongly about creating a Spaceship Earth consciousness. Until at least two to three billion people consider themselves crew members, and take responsibility for their actions, we will have a spaceship with too few crew and too many dependent passengers.

Unfortunately, Number 4 depends on how awake humans are. We are only now reacting in increasingly larger numbers. Can humanity react quickly enough to get ahead of the crisis when we are already so far behind its unfolding?

Growth Economies

Capitalism and the larger category of Growth Economies have existed for well over two centuries. It's all we know. It is our only defined economic reality. We have been conditioned by it and live immersed in it. It is now clear to many that Growth Economies, powered by fossil fuels and with the sole goal to produce and consume more, is the real threat to society and civilization. Financiers, pundits, climate scientists, religious leaders and activists are increasingly reproving our global economy for the harm it is doing to our environment and our collective quality of life.

The reasons that elected officials have moved much more slowly than the people is that they are loath to criticize Capitalism. That is one of our motives in creating the vision of a Finite Earth Economy... as a name for a new Capitalism. In the short term, moving to a Finite Earth Economy will totally disrupt the current metrics listed in Chapter 8. In the long term, it can become the most secure and long-lasting version of Capitalism possible.

The reality of the Shift Age, a concept developed by this author 12 years ago, is that almost everything in this 25-30 year period (roughly 2005 to 2030/35) will change. That the reality we have known until this century and the reality we will live in the 2030s will be fundamentally different. Moving from Growth Economies to a Finite Earth Economy is a major part of this shift, and a necessary component to avoiding oblivion and having a chance at utopia.

Given the almost infinite dynamic variables referenced above, and of course the uncertain unfolding reality of the planet's reaction to what we have done, it is impossible to state definitive timelines and milestones with accuracy. As a futurist, my most important goal is to be accurate in my forecasts of the future. I have been consistently successful in this regard. The problem with that relative to this book is that I have been forecasting catastrophes in the climate change area for decades. Unfortunately, this catastrophic forecast is happening. So, what follows are not forecasts, but suggestions of timelines and milestones that need to be met for a mostly formed Finite Earth Economy to be in place by 2030. We miss them at our peril, the degree of which we can only project. The more we miss these timelines and milestones, the more at risk we place the future of civilization.

NOTE: As you read what follows, please keep in mind the landing of a man on the moon in July 1969. When President Kennedy laid down the vision of "successfully landing a man on the moon and safely bringing him home by the end of the decade" the U.S. had not even put a man into orbit around Earth. (Please reread the quote at the top of Chapter 11). So, some of the milestones and timelines that follow are based on creating things that do not yet exist, particularly in the political realm, and the need to have nations work in tight collaboration on critical global goals and guidelines.

Timelines

In Chapter 9 we suggested some of the major economic actions to take first. First means now. So here is the ideal timeline from now to 2030. Please remember that my estimate is that for every day humanity is not actively moving toward a Finite Earth Economy we will add an additional one billion dollars of economic cost globally going forward. Knowing that this trilogy of book is being published in the second half of 2019, I am making the assumption that the effort starts not today but on January 1, 2020.

1/1/2020 - 12/31/2020

1. In the U.S. and the top ten GHG emissions countries, implement a Carbon Fee and Dividend mechanism. Pass legislation, set up the apparatus to oversee the processes, research and collect all relevant current data, and prepare citizens for the reality of this starting in 2021. NOTE: the unfortunate reality in the U.S. is that unless two thirds of both houses of Congress vote for this, President Trump will veto it. Other countries must move as quickly as they can to compensate.

2. Implement environmental and climate change curricula at the K-12 level and in higher education as soon as possible. Some countries are already doing this.

3. Pass legislation to set up mandatory national service to be focused solely on the climate crisis. If there is any country (most likely the U.S.) that cannot implement at the national level, states and cities should lead with some model programs (policy details in Book 3).

4. Set national deadlines for requisite percentage of EVs on the roads of each country. For example, Norway is already well into the double digits of EVs on the road. This will need across the board support from government and industry with subsidies at the buyer level. An example would be that a country informs its citizens that by 2025, 30% of all cars on the road must be EVs. This will vary by country based upon the percentage they start at and the private sector's ability to ramp up production. The goal is no less than 50% globally by 2030.

5. The same as #4, but with a slightly longer-term target on the ramp up and deployment of Autonomous Vehicles.

6. Nations need to set up and fund national and global competitions to create Atmospheric Cleansing and Carbon Capture technologies that can be scaled up through the decade. The goal should be to start small and scale as fast as possible. The two metrics will be the number of gigatons extracted from the all levels of the atmosphere each year, and the CO_2 PPM in the atmosphere. The respective goals by 2030 are 800 gigatons or less of resident CO_2 and less than 325 PPM.

7. Start to transform Agriculture as discussed in Chapter 11 and Chapter 13. This will be a decade long effort but must start in 2020.

8. By the end of the year, have satellites orbiting Earth that can, on a daily basis, monitor the emissions of every country, or at least the top ten GHG emissions countries.

9. Fund the deployment – private and public – of intelligent sensors globally to get daily/weekly/monthly reports on iceberg melt, sea level rise, emissions at the city level, land

use, deforestation, species extinction and any other data climate scientists think relevant. If the line that "Data is the new oil" is adopted, then capturing data globally will help us get off of oil.

10.   Set up a global council to work with media to publicize, as frequently as possible, how well humanity is doing decreasing GHG emissions. Starting with monthly Emissions Announcement, and becoming more frequent as the technology allows. This solves a major problem relative to climate science, in that the available data is rarely current.

11.   URGENT AND EXTREMELY IMPORTANT: a global entity, perhaps the UN, publishes to the world the metrics set forth in Chapter 10. Media must support this effort in getting the word out and explaining what each metric measures, and why it's important. If at least two to three billion people understand the 2030 goals, there is a much better chance of achieving the goals.

2021

1.   This is the first year humanity can look back at and assess progress on the annual targets set forth in Chapter 10. Given the current lag time in quantified measurement, in 2021 we look at the results from 2020 on:

>   A.   Developing an accepted tight dollar range on the social cost per ton of GHG emissions.
>
>   B.   Moving Earth Overshoot Day back to August 4 in 2020.
>
>   C.   Percentage of global energy that is fossil fuels (76% in 2020).
>
>   D.   Annual global GHG emission for 2020 at 37 gigatons or less.

      E. Annual measurement of C02 PPM in the
            atmosphere at 416 or less.

2. Implement Carbon Fee and Dividend in as many countries as possible starting at $40/ton.

3. Implement wide ranging changes to nation state tax codes. (This will be laid out in Book 3.)

4. States/provinces and cities move on their own to implement policies that result in reducing carbon and issue guidelines for energy sourcing, transportation, waste generation and possibly taxing consumption.

5. Use satellite data to measure daily GHG emissions for the top ten emission countries and the top 10 wealthiest countries per capita. There will be some overlap here, but the point is that the wealthy must compensate for the poorer countries inability to pay their way.

6. Continue public/private funding and development of "moonshot" technologies such as atmospheric cleansing, carbon storage, next gen nuclear, Space-based Solar Power and all transformative developments in energy, transportation, agriculture and any "big" ideas to be deployed later in the decade .

7. Have as many countries as possible, certainly the top ten emitting ones, implement massive national service corps that will plant trees, clean up waste, restore damaged areas of nature and do anything else that lowers emissions and accelerates the conversion to clean energy. Suggestions on this in Book 3.

8. National information campaigns – tied in with education curricula mentioned in 2020 above – to explain the need to move to a more plant-based diet. Everyone should be somewhere on the vegan / vegetarian / "reducetarian" spectrum. This will become easier as a wide variety of plant-based foods become more available.

9. Governments everywhere and at all levels STOP all subsidies to anything that results in emissions (e.g. the fossil fuel industry, the livestock and industrial farming industries), and redistribute them to clean energy technologies and to the necessary remaking of the agriculture industry.

2022

1. Repeat #1 from 2021 with more frequent measurements of emissions for all countries given satellite technology in place. All data immediately announced by the global entity mentioned in #11 in 2020. Data on a weekly and eventually daily basis is the goal. Humanity cannot move toward the numerical goals stated in this book without current data. So all data points that scientists need, and policy makers want, must be widely and immediately available. This must be the real data collected, not what any nation state leader may prefer to announce.

2. Create new metrics based upon the speed of climate change and what is going on at the time. Historic climate change forecasting has over and over been too conservative. Climate scientists are typically surprised by how much faster things are happening than expected.

3. Begin industry by industry measurements of emissions. This can begin this year because of the monitoring systems in orbit and on the ground already in place.

4. Nations and states/provinces announce laws that will go into effect by or before 2025 to regulate and motivate citizens on the move to EVs, the reduction of meat consumption, the rapid transition to clean energy, zoning changes moving to grey water systems, etc.

5. This year is when the externalities of carbon, nature and all earth resources must first be included in corporate financial statements.

6. Implement carbon footprint-based tax codes in every country that taxes its citizens. (More in Book 3)

7. In coordination with #6, governments, nonprofits, educational institutions and all media collaborate on massive initiatives to educate all people about how to reduce their carbon footprint. Tax returns will largely be automated, but tax accountants will need to be well-versed in carbon footprints and offsets.

8. Given the likely issues with governments moving quickly enough, the public must take action wherever governments lag behind. The Finite Earth Economy requires an informed and activist citizenry.

2023

1. Review of the metrics set forth in #1 in 2021. This hopefully has become a standard operating procedure for crew members on Spaceship Earth. We all understand the importance of these numbers and hold each other accountable.

2. At this stage it is expected that social pressure will start to kick in. Examples would be the questioning of people who totally disregard the climate change reality. People who drive big ICE vehicles, live alone in a big house, or any activity that runs counter to the collective need to rapidly lower GHG emissions.

3. Governmental partnerships with the private sector to research, develop and rapidly deploy any and all technologies that accelerate the transition to clean energy.

4. Media outlets report daily on all progress or lack thereof. This becomes a normal part of news reporting just like the weather, traffic and sports. All print and electronic media now include daily updates on the metrics of our macro and micro numerical goals. And it becomes part of the core news reports, with a major story every day.

2024

1. Aggressively move toward the 2030 goals on all metrics.

2. Daily reporting of #1 at global, national, state/province, city and neighborhood levels.

3. Top 10 emitting countries that have implemented a carbon fee and dividend, and carbon footprint tax codes, pressure lagging top 10 countries to hold up their part of the bargain.

4. This is the year where we reach the tipping point within global societies and cultures… it is the default for people to fully engage in successfully reaching our metrics. The Finite Earth Economy is understood and fully adopted.

5. Regulations governing corporate environmental messages are enforced. If a corporation (or any business for that matter) does not have a 100% carbon neutral footprint, they will not be allowed to market themselves with any green messaging. They may even be required to include a warning label similar to the ones on cigarette packages (e.g. "This company does not have a carbon neutral footprint. Use of its products is harmful to the health of all living creatures.").

2025

1. This is the deadline year for all nations to have preliminary plans for Strategic Retreat. There will be growing refugee migrations due to sea level rise, droughts, wildfires, etc. Treaties will need to be signed where a nation agrees to accept refugees from another country, potentially with some sort of compensation process included. More on this in Book 3.

2. Technology enabling frequent, real time measurements of all metrics is deployed. They feed public and private initiatives to incite and inspire all necessary actions.

3. Population management is broached. Many first world nations are already in a negative population growth situation. All countries that still have positive population growth will be incented to address it via education and birth control.

2026-2029

NOTE: This is seven years from when this book is being written in 2019. It is difficult to set specifics. It all depends on what has been accomplished over the previous seven years, and what has not. It would be specious to dictate the specific actions that will be needed at this time.

1. Continuation of all things listed above plus new actions based upon where we are relative to attainment of metrics by 2030.

2. By 2026 all drawdown, carbon capture, atmospheric cleansing and new transformative energy sources and delivery systems are being widely used around the world, particularly in the 10 top emitting countries.

3. Nuclear waste from old nuclear reactors has been successfully and safely sequestered.

4. Ongoing efforts to transition completely from fossil fuels to clean energy are in full operation.

Please note that we have consistently stated "by 2030" or #2030. Specifically, this means by 12/31/2029. If humanity is lagging behind our goals, this will be the year to do whatever it takes to get to 2030 goals. The deadline is looming. All hands on deck... a final push to do whatever it takes.

Takeaways

1. It is essential to have a global mobilization on all fronts.

2. The top 10 polluting countries must take the lead and must be held accountable by the rest of the world.

3. All metrics set forth in this book – and others that will be adopted – must be accomplished by the suggested timeline above.

4. Daily measurements of all metrics are taken and communicated to all of humanity.

5. Urgency must be internalized in all levels of society. Some nations will move more quickly than others. Citizens will push their political leaders to meet their national commitments.

Chapter Thirteen

Design and Redesign

"Wasteful consumerism is a contemporary evil. We need to get back to economic, plain, simple and useful design."
– Terence Conran, legendary British designer

"We all live down-wind, down-stream, and down-time from everybody else. We need to think, design, and operate in a systems model – a holistic context of dynamic equilibrium." – Houle & Rumage, This Spaceship Earth

"If it can't be reduced, reused, repaired, rebuilt, refurbished, refinished, resold, recycled or composted, then it should be restricted, redesigned or removed from production." – Pete Seeger

"We stand at a new threshold, and the key questions of our day are those posed by the intersection of design and life itself." – Bruce Mau

"We have entered the Golden Age of Design." – David Houle 2016 AICAD keynote

Much of what we need to do to successfully slow global warming and alleviate the climate crisis is related to design. Design and redesign are crucial components of the Finite Earth Economy. The 20th Century was left brain, the 21st century is right brain. The right brain leads with design, creativity and humanity. The left brain engineers and executes on the right brain design/redesign thinking.

So many of the design decisions made every day have a climate implication: each one can help promote a low-carbon future that doesn't rely on fossil fuels. Those who create the products and built environments of everyday life – from mechanical engineers to architects – have an important role to play by designing for climate change.

Many of the issues humanity is facing today can be effectively addressed with design thinking. We are living in a world primarily designed for a population of 2.5 billion. In 1950 there were 2.5 billion people alive. Today in 2019 there are approximately 7.7 billion!

100 years ago, we were less than 2 billion. There were seemingly unlimited land and resources. The world had begun the great migration from country to urban areas as the Industrial Revolution grew to affect every facet of life. Suburbia had yet to be developed. The only communications towers were for radio… broadcast TV, cable TV and cellular technologies had yet to be invented. Interstate highway systems in all developed countries had yet to be conceived.

Humanity must redesign its thinking around population. Until recently, population growth was perceived as a good thing. Largely, this was driven by three forces. First, in 1900 a "normal" American family had 5 children. Often one or two died from childhood diseases. This is no longer the case. Improving medical science and sanitation has largely extended life expectancy. Second, religions have always wanted to populate the planet with offspring (therefore increasing the number of their "faithful"). There is no major legacy religion I know of that limits population growth as a part of its belief systems. Just the opposite. Third, an expanding population has a direct correlation to GDP. More people means more consumption, which means more production and a higher GDP

In Book 3, we will examine all levels of society: global, nation state, state/province, city, local and personal to determine what each must do to transition to a Finite Earth Economy by 2030. These include devising entirely new tax codes, policies and regulations. All of the existing ones have largely been put into place in the post-WWII era. They were designed for a fossil fuel economy and are no longer valid. Updating some, completely replacing others and adding new ones will require truly creative design thinking.

Outdated Infrastructure

During the 20th century, technologies invented at the end of the 19th century and the beginning of the 20th became institutionalized. The internal combustion engine became institutionalized as Transportation. Machinery created the industrial stage of agriculture. The factory became the model for K-12 education. Microwave towers formed the electronic communications systems. This became our landscape and lifestyle until the end of the century when the Internet, digitization, Moore's law, satellites and a population explosion worked together to make it obsolete.

Here we are in the first quarter of the 21st Century with much of the world built out, paved over, and with ever more overcrowded mega-cities. Humanity is in the process of moving from 55% of 7.7 billion people living in urban areas to a projected 70% of 9 billion living in them by 2045, a total increase of 2.45 billion from 3.85 to 6.3 billion. This means that in the next 25 years about 100 million new inhabitants will move into urban areas annually.

So, we have to redesign the much of the built environment. We have to design what the 21st century urban environment will be. We must make design/redesign decisions on transportation, housing, and highest use real estate retrofitting. A prime example would be to take the ever-growing numbers of vacant big box stores and recast them as indoor vertical gardens, providing multi-crop yields of organic produce that is locally grown.

We have to completely redesign the 20th century electrical grid to the more decentralized, distributed energy landscape of this century. We need to redesign all forms of industry, transportation and daily life from being powered by fossil fuels to being powered by clean, non-polluting energy sources. It's a gargantuan (and exciting) design effort.

We interact with infrastructure all the time – from the pipes that bring water each morning to our showers, to the roads we drive daily to work, to the garbage collections services that haul away our waste. Much of the time, it can be easy to forget that these working systems even exist. They simply work. It's when infrastructure fails that real challenges can arise: from the physical infrastructure we depend on, to the infrastructure that underpins modern manufacturing and supply chains.

Imagine a scenario where infrastructure can adapt and respond. Let's use water pipes as an example. Imagine pipes that could adjust to fluctuations in temperature, water flow, and pressure to prevent damage. Designing for repair is key: the most sustainable product is often the one that lasts the longest. Imagine that the pipe could even heal itself once broken or burst. While still in its early, largely R&D stages, 4D printing technology can potentially enable this type of infrastructure functionality, saving effort and costs related to business operations and parts maintenance.[1]

"As for retrofitting the twentieth century, the **what** *is becoming clear. The* **how** *is beginning to be discussed. There is developing urgency around the* **when.** *The* **who** *is all of us. (And the* **where** *of course is our planet.) Retrofitting the twentieth century will be a signature effort of the Shift Age as almost everything is now in relative rates of shift. Historians in the year 2100 will look back on the 2010-2020 decade as when this retrofitting began in earnest… just in time."* – David Houle "Entering the Shift Age" 2012

Strategic Retreat

The climate crisis is creating refugees due to droughts, floods and sea level rise. In the near future, the numbers of refugees will increase exponentially. We will have to design – on a global scale – how to move and resettle tens of millions of climate change refugees. The greatest relocation project in human history.

Whole Systems Problem Solving

Designers and engineers who use the process of whole-systems problem solving consider the relationships among complex systems, instead of focusing on individual parts of systems. This is important because challenges such as climate change represent a set of interconnected issues that can't be solved in isolation. By taking a big-picture view and considering the whole system, the most important opportunities often arise, and can be incorporated, early in the process.

This type of systems thinking leads to holistic design. Holistic design sees a problem or a system as part of a larger, interconnected whole. Given that Nature is the most complex, integrated, holistic design humans have ever experienced, it is obvious that systems thinking and holistic design are core to the redesign of how we live as we move to a Finite Earth Economy.[2]

Biomimicry and Cradle to Cradle Design

The Finite Earth Economy will include a different understanding of consumption. With the goal being to produce abundance, but to do it without waste or emissions. This can be done. Mother Nature produces abundance all day every day with no waste or toxicity. She accomplished this via 3.8 billion years of trial and error R&D.

Could we model our own product design on Mother Nature's? A tree produces thousands of blossoms in order to bear fruit, yet that abundance isn't wasteful. It's safe, beautiful and highly effective. The blossoms fall away to biodegrade and enrich the soil. If not picked, the fruit is eaten by birds and seeds are distributed to produce more trees. Products might be designed so that, after their useful life, they provide nourishment for something new – either as biological nutrients that safely re-enter the environment, or as industrial nutrients that circulate within closed-loop production cycles, feeding the manufacture of other goods.

Mother Nature has her own waste disposal solutions. In nature nothing is wasted. Waste not, want not. All things that were once alive eventually become part of the earth again, returning usable resources. So why not study how nature produces abundantly without waste?

Bill McDonough, the architect and designer, has formulated a design principle which he lays out in his book *Cradle to Cradle*. He demonstrates that "reduce, reuse, recycle" is based on a 'cradle to grave' design model that dates to the Industrial Revolution. This process casts off as much as 90 percent of the materials it uses as waste, much of it toxic. Mr. McDonough challenges the notion that human industry must inevitably damage the natural world.[3]

There's an established science that analyzes nature's ideas and adapts them for human use. It's called biomimicry. So far, most biomimicry breakthroughs have been designed and deployed to build better products – lighter, stronger, faster, etc. What if biomimicry scientists (biomimicists?) tackled Bill McDonough's closed loop, zero waste, cradle to cradle design challenges?

By all means, we should continue to reduce, reuse and recycle; as well as develop sources of clean energy and green technologies to mitigate and adapt to climate change. But those are half measures. In the Finite Earth Economy, we want to continue (and expand) our current quality of life, so we should borrow design and production techniques from Mother Nature.

Agriculture

Agriculture is an industry that is going to be thoroughly redesigned. For example, as ever more people move to urban environments there will be a growing need for food grown locally within cities to save CO_2 emitted for transportation. This both reduces the carbon footprint of long-range transport and also leverages the 'food to table' movement... resulting in fresher, more nutritious food. The collapse of physical retail in the last decade has created countless numbers of vacant big box stores. They can be converted into vertical gardens that have four to six crop yields per year.

Farmlands throughout the world need to be redesigned for the Finite Earth Economy. Here are some types of agricultural redesign:

Intercropping or companion gardening intersperses different varieties of crops and plants that are mutually beneficial to each other. For example, planting nitrogen fixing plants between rows of plants that uptake nitrogen. Or planting flowers amongst the crops to ensure an abundant supply of pollinators year round. Or planting "trap crops" which lure pest species away from the cash crops.[4]

Staggered planting, whether at indoor vertical farms or intercropped farms, staggers when a crop is planted. Instead of planting an entire field at once, one section is planted every couple of weeks to let the soil rest and lengthen the harvest. Indoor farms can do staggered planting year round. If there's a glut in the market for a certain crop, the farmer can plant something else.

By moving agriculture into the fabric of the city, you eliminate most of the transport cost and can allow the crops to ripen on the vine, picking and delivering them to stores and restaurants on the same day. Another benefit of growing in the city is that food waste is close to the farm sites. Composting becomes convenient and cost effective. Food waste does not end up in the landfill producing methane. And the need for commercial fertilizers is minimized.

Urban agriculture is much less mechanized so there's no need for the cost and debt of large farm machinery.

Redesigning the Economy

The biggest transition, alongside the massive shift away from fossil fuels, will be the redesign of Growth Economies to Finite Earth Economies. An evolution from a "growth first" economy to an "earth and humanity first" economy will require a vast amount of design thinking, followed by an unprecedented level of economic reengineering. It will be one of the most intense and complex design/redesign undertakings in history.

In this book we have discussed all the current and new technologies that we must deploy as we move to a Finite Earth Economy. Here are just some of the energy and technology redesign transitions:

- developing distributed energy models to coexist with the current grid model,

- redesigning the refueling infrastructure replacing and/or supplementing gas stations with electric charging stations,

- retrofitting and redesigning outdated 20th century real estate (especially retail outlets including malls and big box stores),

- moving from a society designed for the one car per person ownership model to one where there are ever more short-term rentals, ridesharing apps and autonomous vehicles,

- the continuing evolution from physical to digital and how that affects housing, schools, entertainment and shopping,

- the design and deployment of "moon shots" such as Space-based Solar Power and Atmospheric Cleansing technologies.

Relative to the new societal and energy metrics needed to measure our progress to a Finite Earth Economy, we will need to monitor all forms of greenhouse gas emissions, sea level rise, waste reduction, the decrease in consumer spending on non-food and housing items, and the lessening of carbon footprints. This will require the design of a monitoring infrastructure that is effective, unobtrusive and that can be easily extended as new metrics are introduced.

In Book 3, we will suggest how design teams can be mobilized at the planetary, national, state/province, city and neighborhood levels. At the core of all design and redesign projects will be several tenets:

- approach each design problem from a holistic perspective using systemic design principles,

- lower greenhouse gas emissions as quickly and completely as possible in every facet of our economy and society,

- restore as much of the natural world as possible through clean-up, conservation and restoration,

- create direct linkages between what people do and the consequences they unwittingly create,

- replace quantity with quality,

- look to Mother Nature for elegant design solutions,

- design all products to have a long life of utility, and to be easily disassembled and reintroduced into the production life cycle when their useful life is over.

In short, the design and redesign of the global Growth Economy to a Finite Earth Economy is the single biggest design project humanity has ever faced. And design thinking will inform all of our decisions in implementing the Finite Earth Economy

We are up to the task, but our timeline is short, and civilization is at stake.

Takeaways

1. Here in the 21st century, right brain thinking must infuse humanity into the left brain designed 20th century. Designers will collaborate closely with engineers and scientists.

2. Humanity must retrofit much of the built environment and infrastructure that was created within a Growth Economy

mindset, and redeploy it under Finite Earth Economy thinking,

3. Design thinking will be at the center of the coming Strategic Retreat [Book 3] which will be the largest human relocation project in history.

4. Whole systems problem solving combined with holistic design is at the center of all that we need to do.

5. Biomimicry and Cradle to Cradle design must be further developed, expanded and deployed as central design concepts for a Finite Earth Economy.

6. Agriculture, the leading producer of GHG emissions must be completely redesigned. How we grow, where we grow, what we grow, and what we eat will all be redesigned.

7. Moving to a Finite Earth Economy means completely redesigning the global economy, the biggest systems thinking and holistic design project in history.

Moving to a
Finite Earth Economy

Crew Manual

Book Three

Global to Personal

Chapter Fourteen

At Every Level – Global to Personal

"We are not going to be able to operate our Spaceship Earth successfully, nor for much longer, unless we see it as a whole spaceship, and our fate as common. It has to be everybody or nobody." – R. Buckminster Fuller

"The illiterate of the 21ˢᵗ Century will not be those who cannot read or write. Instead it will be those who cannot learn, unlearn and relearn." – Alvin Toffler

"If man chooses oblivion, he can go right on leaving his fate to his political leaders. If he chooses Utopia, he must initiate an enormous education program - immediately, if not sooner." – R. Buckminster Fuller

We have to stop focusing on artificial dualities and past battles and move forward at every level. We set up global metrics for the move to a Finite Earth Economy in Book 2. Here in Book 3 we break it down to the general categories of how humanity thinks of itself and how humanity has been organized. These are nation states, corporations, states and provinces, cities, neighborhoods and individuals.

There are entities such as the UN, NATO, EU, the World Bank, and the WTO. All these organizations have pan-national status and often great influence. That said, they are run as member organizations and the members are nation states. The nation states vote and retain full sovereign decision making. In this century we are moving to a more globally integrated discourse and economy. This will ultimately evolve to some sort of global governance or at least an authority agreed upon issue by issue. The road that humanity must travel in the next 10 years will require collaboration and cooperation across disciplines well beyond anything humanity has ever attempted.

This is the first time that every living thing on Earth has the same common enemy: how humanity lives on Earth.

It is past time to save ourselves from ourselves.

Nations are currently the highest level of political and governmental organizations. The current problem is that nations think nationalistically, not globally. As a futurist, for the past 15 years I have stated that the 21st century will be known as the century when the concept of the nation state will remain, but will no longer be the premier authority. We are becoming ever more connected and integrated globally.

The key to moving to a Finite Earth Economy is to have cooperation and integration at all levels of humanity: nations, corporations, states/provinces, cities and neighborhoods. We live on one planet, though we have many nations and religions. We need to realize that we are in a critical moment for the survival of civilization, and cannot allow our perceived differences to put all of humanity at risk. Relative to moving to a Finite Earth Economy by 2030, we must all think and act as crew of Spaceship Earth above all other self-identifications.

This means, to the point of the Toffler quote at the top of this chapter, that we must unlearn how our Growth Economies work, and learn how to create and live in a Finite Earth Economy.

Below are two pictures of Earth. When making presentations about climate change I put up the first image and ask the audience, "What this is a picture of?" Everyone says Earth! Then I put up the second image and say, "No, this is Earth. The first one is a picture of nation states. Humans are the only species to have political boundaries."

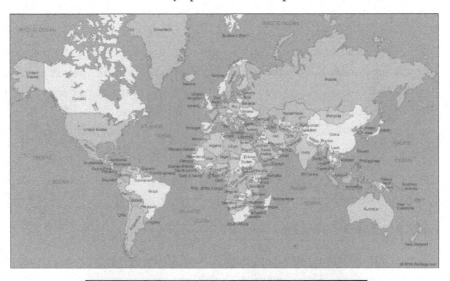

Image Credit: NASA

Book 2 was global in perspective with timelines on what humanity will need to do to successfully move to a Finite Earth Economy by 2030. This book starts at the nation state level and works down to the personal level to identify what must be done at each.

Think of this third book of the trilogy as the answer to the questions:

What can we do? What can I do?

Concurrent and Interrelated

We must begin at all levels, now. There is no level of human social organization that has the luxury to wait. At the global level down to the individual level, all activities must be infused with a full-blown sense of urgency. We have delayed and acted incrementally for decades. Now, we only have one decade left to reach our goal of living in a Finite Earth Economy by 2030.

The fork in the road: Utopia or Oblivion

Utopia is not certain, but is a true possibility. It can be reached if we think, coordinate and act with an all-inclusive systems approach as R. Buckminster Fuller stated in the quote at the top of this chapter. All is interconnected on Earth. All levels of human society are interconnected. All of nature is interconnected and we are a part of the natural world. A single person can take personal actions, can participate at the neighborhood level, at the city level, and can vote at the state and national levels to integrate her efforts.

Being an aware person who consciously lowers her carbon footprint, lowers her consumption, and changes behavior from being purely selfish and self-indulgent to acting for the greater good is laudable… but it does not absolve her from participating in neighborhood activities, or from putting pressure on local authorities, or from voting in all state and national elections.

"What else can I do? Who else can I help and influence? Who needs to get out of the way? Who do I need to support politically?"

Oblivion as manifested by a collapse of civilization will happen if we do not take urgent action to move to a Finite Earth Economy. We have free will. It is up to us to decide on the path we are going to take. Do we band together and take a collective activist approach to 2030, or not? Remember, the three possible paths to 2100: the catastrophic (business as usual), the risky (protest, recycle, complain and condemn), or a highly probable success (full blown rework of how we live on Spaceship Earth). This third book of the trilogy suggests actions at all levels to make our successful outcome highly probable.

If you have been a long-time environmentalist, someone who has been concerned about climate change for years or decades, you can no longer use that as an identity differentiator. You must do more. You must reach out and educate those who are still oblivious. Inspire them to take action too. We must ALL become active crew members on Spaceship Earth.

New metrics and timelines were detailed in Book 2. In this book we show how we change and how we meet our milestones on this critical timeline.

There is much to do and no time to wait.

Takeaways

1. Actions to move to a Finite Earth Economy by 2030 must be taken concurrently at every level of society.

2. We have a choice: Utopia or Oblivion. Utopia can only occur if we mobilize aggressively at every level of human society.

Chapter Fifteen

Strategic Retreat

"The Shift Age will be the greatest age of migration in history. A greater number of humans will migrate at some time of their lives than in any prior age. Some of this will be due to the fact that the human population is larger than ever before. This means that the sheer number of us will be a factor. In addition, the percentage of the total population that will experience short to long term migration will increase as well." – David Houle, The Shift Age, 2012

Strategic Retreat is one of the most critical challenges we face as a consequence of climate change. Along with Population (next chapter), it's a topic that hasn't been covered in much detail in the climate change literature. This is largely due to the fact that, until recently, most books and media reports about climate change were focused on substantiating that climate change is "real" and that we must reduce carbon emissions.

Climate change is HERE and NOW, and we have entered the fight or flight stage. It is clear that we are facing a challenge the scale of which is unprecedented in human history.

Strategic Retreat is the planned and managed relocation of millions of people triggered by extreme weather, droughts, floods and Sea Level Rise [SLR]. Mass migrations as a byproduct of climate change are inevitable. To minimize social strife and human suffering, they must be thought through and planned for. Sometimes it is called managed retreat, but that implies having to manage it, which is the second stage. The first stage is to strategically plan as much as we can... starting now!

The level of GHGs in our biosphere has never been higher in the time man has been alive on Earth. The resulting global warming and our current climate crisis are new phenomena. We have no reference point for dealing with them. How do we prepare for the relocation of tens of millions of people in the decades ahead? There are some mass migrations in history we can reference, and we will discuss them later in this chapter.

Extreme Weather Events

As we all are aware, extreme weather events are happening with increasing frequency. "500 year" floods and droughts are occurring every few years. Massive cyclones and hurricanes are devastating large swaths of land all over the globe. Forests and brush lands covering millions of acres are burning, destroying entire towns and small cities.

It is one thing to rebuild a home after a flood destroys it, with the expectation that the next flood will not occur for decades. It is a completely different thing if it can be forecast that that home will be flooded again and again every few years. At some point, people give up. They decide to move elsewhere. Wealthy people can absorb the loss and buy a new home in a safer location. Poor people don't have the wherewithal to do that. They become displaced.

It is impossible to determine exactly the number of people who will be forced to migrate due to climate change. There is no historical reference for a migration this large. The range of estimates is 10 to 30 million climate refugees between now and 2030, and 50 to 140 million by 2050. I think that it will start slowly in the early 2020's and accelerate by mid-decade.

As more areas of our world become uninhabitable, large refugee migrations will become more frequent. How will we handle them? Where will they go? Who will pay?

Sea Level Rise

Just how much SLR we will experience depends on too many factors to forecast with accuracy. Current projections are all over the place. We know SLR will occur at different rates and at variable levels depending on location and the speed at which land ice is melting around the world. We know that the oceans are rising at ever faster rates. The pace of global sea level rise nearly doubled from 1.7 mm/year throughout most of the twentieth century to 3.1 mm/year since 1993.[1]

Note: For those new to climate science, a brief explanation on the physics of SLR. Sea level is rising for two main reasons: glaciers and ice sheets are melting and adding water to the ocean, and the volume of the water is expanding as the ocean warms.

To give you an example of the amount of water stored in land-based ice and glaciers, the Greenland ice sheet is 10,000 feet thick in places and contains enough ice to raise sea levels 23 feet (7 meters).[2]

While this level of SLR will not happen in this century, it is clear that humanity has taken a planet at equilibrium and triggered a process that will result in significant changes to many characteristics of the biosphere.

In Chapter 7 of Book 1, we charted three general paths forward for humanity. The first was catastrophic and based upon a deadline of 2100 to complete the transition away from burning fossil fuels. In recent years, research and extreme weather events have caused the consensus deadline for the end of fossil fuel use to be moved up to 2050. The best probable outcome is one where humanity moves to a Finite Earth Economy as much as possible by 2030. This means cutting fossil fuel use to 30% of all energy by 2030.

What follows are some charts of locations around the world that are vulnerable to SLR. Just below is a table with forecasts based upon the three paths forward. There are too many variables for anyone to make guaranteed forecasts, but we've looked at a comprehensive number of scientific estimates to come up with what we feel are numbers as solid as possible.

Global Sea Level Rise Forecasts for the Three Paths/Scenarios

	2050	2100
Catastrophic	4+ feet	9+ feet / 3+ meters
Risky	2+ feet	5+ feet / 1.5+ meters
Best Probable	1+ foot	3+ feet / 1 meter

Recently the World Economic Forum stated:

The World Economic Forum's Global Risks Report 2019 estimates 800 million people are vulnerable to a rise in sea level of 0.5 meters by 2050. That includes 70 percent of Europe's largest cities, 19 African cities with populations over one million, and 78 million Chinese in low-elevation cities, a number that grows 3% every year. That 0.5-meter change in sea level — barely half of the lower projection — would result in the loss of about 11% of land in Bangladesh, displacing as many as 15 million people.

Here are some charts showing the cities and countries at greatest risk of SLR.

Rank	Country	Urban Agglomeration	Exposed Population Current	Exposed Population Future
1	INDIA	Kolkata (Calcutta)	1,929,000	14,014,000
2	INDIA	Mumbai (Bombay)	2,787,000	11,418,000
3	BANGLADESH	Dhaka	844,000	11,135,000
4	CHINA	Guangzhou	2,718,000	10,333,000
5	VIETNAM	Ho Chi Minh City	1,931,000	9,216,000
6	CHINA	Shanghai	2,353,000	5,451,000
7	THAILAND	Bangkok	907,000	5,138,000
8	MYANMAR	Rangoon	510,000	4,965,000
9	USA	Miami	2,003,000	4,795,000
10	VIETNAM	Hai Phòng	794,000	4,711,000
11	EGYPT	Alexandria	1,330,000	4,375,000
12	CHINA	Tianjin	956,000	3,790,000
13	BANGLADESH	Khulna	441,000	3,641,000
14	CHINA	Ningbo	299,000	3,305,000
15	NIGERIA	Lagos	357,000	3,229,000
16	CÔTE D'IVOIRE	Abidjan	519,000	3,110,000
17	USA	New York-Newark	1,540,000	2,931,000
18	BANGLADESH	Chittagong	255,000	2,866,000
19	JAPAN	Tokyo	1,110,000	2,521,000
20	INDONESIA	Jakarta	513,000	2,248,000

Table 1: Top 20 cities ranked in terms of population exposed to coastal flooding in the 2070s (including both climate change and socioeconomic change) and showing present-day exposure (Source: Nicholls et al (2007), OECD, Paris)

Chart 15.1

As you can see from the above table, most of the large cities threatened by SLR are in Asia.

There will be tens of millions of people forced to move due to SLR. Where will they go? Will there be compensation for the lost value of their coastal homes? Who will pay for their resettlement?

Here's an example closer to home. Miami-Dade county has 2.75 million residents and is the seventh largest county in the U.S. This county is already suffering the early consequences of SLR. Every month parts of Miami Beach are under water purely due to high tides. It is a phenomenon called 'sunny day flooding' and is now becoming common in Florida and up the East Coast.[3]

Where will these people move to once real estate values start to collapse, their freshwater aquifer is tainted by saltwater, and there is standing water in most of the streets, most of the time? There currently are initiatives either under way or planned to raise city streets and build sea walls. The projected costs are in the billions and these are just partial solutions. What about all the homes and businesses? Who is going to pay to raise them up by several feet? How do you raise a thirty-story condo tower? Will they lose their real estate investments? When will the banks stop lending for construction on parcels at risk for SLR?

Insurance companies (especially reinsurers) have seen this coming for decades. Will they pick up the tab? Will the state or federal governments? Hardly likely.

As a resident of Florida, I have been speaking about SLR and real estate values since 2016. In that year I asked a group of realtors at their annual luncheon, "When does it become fraud to sell beach front real estate?" The remark was not well-received in the room. Fortunately, there was a journalist present, and my speech, along with my controversial question, was on the front page of the local paper the next morning and on the local news that evening.

The most expensive residential real estate is on the beach or on barrier islands. Those are exactly the locations most at risk for SLR. What happens to property values when a majority of residents "get it" and everyone is selling? What happens to beach economies and tourism when the beaches are under water?

Will the U.S. simply wait to see how it unfolds? I used Miami as the example because it is already dealing with SLR. There are several other cities that will need to adapt or retreat (resulting in climate refugees). These are the U.S. cities with the most property, by dollar amount, at risk from sea level rise:

Rank	City	Value at Risk	Share in Risk Zone
#1	New York	$87.3 billion	7%

#2	Miami Beach	$37.6 billion	85.2%
#3	Boston	$35 billion	22%
#4	Honolulu	$23.1 billion	32.2%
#5	Fort Lauderdale	$23.1 billion	61.2 %
#6	Miami	$19 billion	35%
#7	Newport Beach, CA	$16.2 billion	18%
#8	San Mateo, CA	$14.6 billion	41.6%
#9	Hilton Head, SC	$11.3 billion	64.7%
#10	Huntington Beach, CA	$11.2 billion	18.5%
#11	Charleston, SC	$10.6 billion	38.6%
#12	Saint Petersburg, FL	$10.3 billion	30.4%
#13	Miramar, FL	$10.2 billion	81.5%
#14	Virginia Beach	$9.5 billion	17.3%
#15	Jersey City, NJ	$9.5 billion	20.2%
#16	Hollywood, FL	$9.3 billion	45.1%

#17	Cambridge, MA	$9.1 billion	33.4%
#18	Redwood City, CA	$9 billion	30.5%
#19	Hialeah, FL	$7.9 billion	64.5%
#20	Tampa	$7.8 billion	10.3%
#21	Pembroke Pines, FL	$7.6 billion	55.1%
#22	Alameda, CA	$7.3 billion	38.1%
#23	Long Beach, CA	$6.9 billion	5.8%
#24	Mount Pleasant, SC	$6.8 billion	34.2%
#25	Ocean City, MD	$6.1 billion	72.7%
#26	Davie, FL	$6.06 billion	64%
#27	Sarasota, FL	$5.8 billion	18.6%
#28	Pompano Beach, FL	$5.6 billion	34%
#29	Stockton, CA	$5.4 billion	22%
#30	Galveston, TX	$5.05 billion	78.4%

See Footnote 15.4

When will the migrations start? They already have. On average, 26 million people are displaced by disasters such as floods and storms every year.

The UN International Organization for Migration forecasts 200 million environmental migrants by 2050, moving either within their countries or across borders. Many of them will be coastal populations.

Historical Migrations

There are mass migration precedents in our history. Here are a few examples:

Afghanistan to Pakistan	1980s	2.8 million
All Countries to Israel	1948 to 2000	3.6 million
The Petition of India	Late 1940s	10 million
Nazi Expulsion	1935 to 1945	12 million
Stalin's Forced Migrations	1929 to 1952	12 million
African Slave Trade	1400s to 1800s	12 million

These are massive numbers of refugees, and the human suffering that accompanied these migrations is horrifying. Yet they pale in comparison to the numbers of people about to be displaced from their homes due to climate change. It is ludicrous to think that it will work out okay without coordinated thought, planning and policy preparation across the globe.

Legal Issues

There is an interesting legal dilemma regarding all these people who will be displaced. Since there has never been a refugee problem due to climate change, it has not been incorporated into legal definitions. The legal definition of "refugee" means someone escaping persecution for political or religious reasons.

Technically, all the people displaced by climate change do not qualify for refugee status, although their personal losses may be as egregious as war time refugees. All refugee laws (national and international) must be rewritten. The problem is the magnitude of the numbers. We must develop a new refugee status outlining the rights of the tens of millions of people who will have to move due to climate change.[5]

Immigrations Regulations

Immigration restrictions are not inherently racist. Nor is border security. All countries have borders and restrictions. They have to, because they have to make decisions about who can enter their country and who can be a citizen. Nations can't function without such basic laws.

As we write this, the U.S. is experiencing a divisive conflict over immigration. There are those who want to close the borders... and there are those who believe we should accept anyone who wants to enter with open arms. Neither of these is an acceptable solution. What must be done lies somewhere between these polar opposites.

Climate refugees and strategic retreat are not just American issues. Every country on the planet will have a role to play. What that will be, how the rules get made, and how it all plays out remains to be seen. It's a problem and an opportunity. An opportunity for global compassion, collaboration and cooperation. We will see if humanity is up to the challenge.[6]

Strategic Retreat is a huge issue. One that will affect billions of people (if we count those who may not have to migrate, but will be impacted by the issue in one way or another). We must start to work on it NOW, or we will surely experience the tearing of the social fabric globally, and unprecedented social, economic and cultural disruption.

In the end, climate migration experts agree, the real key to tackling this crisis is limiting global warming in the first place.

Please ask yourself: "What else can I do? Who should I help? Who needs to get out of the way? Who do I need to support politically?"

Takeaways

1. Strategic Retreat is one of the most consequential developments of the Climate Change Era. It is also one of the least discussed.

2. Humanity must start to discuss, strategize and plan for the coming migration of over 100 million people in the next 30 years.

3. Humanity must face and accept that extreme weather is the new norm, and take action to manage the risks.

4. Humanity must accept that SLR is occurring and should plan for it, budget for it, and construct infrastructure to manage it.

5. We are currently on the Catastrophic path to 2100. Only by moving to a Finite Earth Economy by 2030 does humanity have a chance to slow warming and thus SLR and extreme weather.

Chapter Sixteen

Population and Family Planning

Those who fail to see that population growth and climate change are two sides of the same coin are either ignorant or hiding from the truth. These two huge environmental problems are inseparable and to discuss one while ignoring the other is irrational." – James Lovelock, scientist and environmentalist

"The raging monster upon the land is population growth. In its presence, sustainability is but a fragile theoretical concept." – E.O.Wilson, biologist

"Everybody with a womb doesn't have to have a child, any more than everybody with vocal chords has to be an opera singer." – Gloria Steinem, feminist, journalist and activist

"We have been God-like in our planned breeding of our domesticated plants and animals, but we have been rabbit-like in our unplanned breeding of ourselves." – Arnold Toynbee, economic historian

"In places where women are in charge of their bodies, where they have the vote, where they are allowed to dictate what they do and what they want… the birth rate falls." – Sir David Attenborough, naturalist

Population and climate change are directly linked. Every additional person increases carbon emissions (the rich more than the poor) and increases the number of climate change victims (the poor more than the rich).

Because individuals in the developed world have the greatest impact per person, people choosing to have smaller families in the richest parts of the world will have the greatest and most immediate positive effect – a vital choice given the urgency to address climate change. Furthermore, reduced emissions as a result of fewer people being born in richer countries allows more economic development in poorer countries without adding to total emissions.

Currently the net daily increase in human population, after subtracting deaths from births, is 220,000 people, or over 150 people every minute (395,000 births a day minus 175,000 deaths).[1] That equals over 80 million more people every year. That is 220,000 new dependent passengers on Spaceship Earth every day, all of whom will generate demand for resources and will cause GHG emissions.[2]

Population growth is one of the most important topics relative to climate change and moving to a Finite Earth Economy. The personal decision of whether to have children or not touches on morality, religion, sense of self, upbringing and long-held beliefs. Those beliefs are of questionable value and relevance in the 21st century. These legacy viewpoints may be why population has not become the primary topic of discussion relative to climate change. Climate scientists and activists may see the world through these beliefs.

To state it upfront: the decision to have a child is the single most environmentally impactful decision any of us can make in our lifetimes.

Having one fewer child will save 58.6 tonnes of CO2-equivalent per year. Here is a chart to put this in perspective:

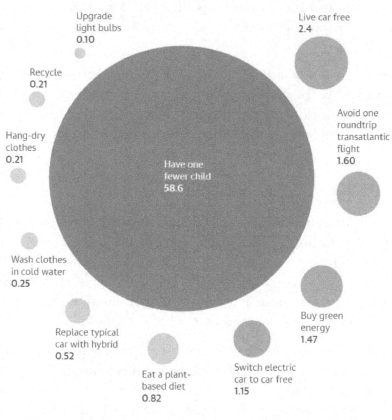

Upgrade
light bulbs
0.10

Live car free
2.4

Recycle
0.21

Hang-dry
clothes
0.21

Avoid one
roundtrip
transatlantic
flight
1.60

Have one
fewer child
58.6

Wash clothes
in cold water
0.25

Replace typical
car with hybrid
0.52

Buy green
energy
1.47

Eat a plant-
based diet
0.82

Switch electric
car to car free
1.15

Chart 16.1 – Guardian graphic | Source: Wynes & Nicholas,
Environmental Research Letters[3]

The figure was calculated by totting up the emissions of the child and their descendants, then dividing this total by the parent's lifespan. Each parent was ascribed 50% of the child's emissions, 25% of their grandchild's emissions and 12.5% for their great grandchild. In other words, the amount of my genetic material in the child is 50%, the amount that's in my grandchild is 25%, and the amount that's in my great grandchild is 12.5%. Divided by the parent's lifespan. So, say the average age of the parent when the child is born is 25, and average life span is 75, so divide by 50.[3]

If an exceedingly committed environmentalist lived a life where she did all of the 10 satellite actions in the above chart, it would result in a savings of 8.52 tons per year. There are few people in either the developed or undeveloped countries that can do all of these. For each baby born there would have to be 7 people completely committed to living a green lifestyle to offset the decision to have one child.

In practically every article or book written about the need to stop burning fossil fuels, there is a chart showing the dramatic increase in CO2 emissions since the beginning of the Industrial Revolution. Here is one that tracks population growth right next to it.[4]

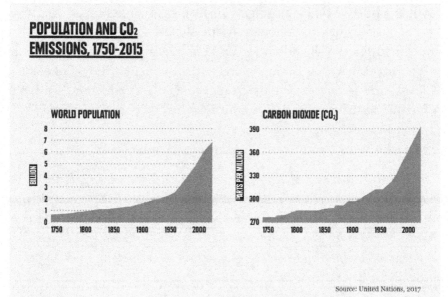

Chart 16.2

CO2 parts per million has a direct correlation with population growth. Human population grew from 1.6 billion to 6.1 billion people during the course of the 20th century. It took from the dawn of humanity for population to reach 1.6 billion in 1900. Then it shot to 6.1 billion in 2000. During the same 100 years emissions of CO2, the leading greenhouse gas, grew 12-fold. The two drivers of this increase were the growth in population combined with a dramatic increase in per capita emissions.

We are currently at 7.8 billion people on Spaceship Earth. Worldwide population is expected to surpass nine billion by 2050, and reach between 10 and 11 billion by 2100. Given the above charts, it is clear that to continue both our current per capita level of fossil fuels use, while adding another three billion people in this century, is the Catastrophic path we laid out in Chapter 7 of Book 1.

Until such time as humanity dramatically lowers GHG emissions, we must explore how humanity could limit population growth. A 2009 study of the relationship between population growth and global warming determined that the "carbon legacy" of just one child can produce up to 20 times more greenhouse gas than a person will save by living a fully environmentally friendly lifestyle. Each child born in the United States will add about 9,441 metric tons of carbon dioxide to the carbon legacy of an average parent. The study concludes, "Clearly, the potential savings from reduced reproduction are huge compared to the savings that can be achieved by changes in lifestyle."[5]

One of the study's authors, Paul Murtaugh, warned that:

"In discussions about climate change, we tend to focus on the carbon emissions of an individual over his or her lifetime. Those are important issues and it's essential that they should be considered. But an added challenge facing us is continuing population growth and increasing global consumption of resources… Future growth amplifies the consequences of people's reproductive choices today, the same way that compound interest amplifies a bank balance."

I know first-hand how sensitive this topic is. During 2016 and 2017, I gave dozens of speeches around the country based upon my book, "This Spaceship Earth". In these presentations, I would first set the stage with all the charts, data and visuals about global warming and climate change. Then I would explain why humanity needs to evolve its consciousness from being unaware passengers to being conscious crew members on Spaceship Earth. Finally, I would end with a prescriptive list of actions that "we" (humanity) must take. The concept of population planning was always on that list.

When I gave some 50 presentations of "This Spaceship Earth" in 2016-2017 this was the one point that had some people squirming in their seats. The concept of population planning makes people feel uncomfortable, guilty or defensive. Totally understandable. Obviously, audience members have been children, most have had children, and many had grandchildren. It became clear to me that having kids, and having as many as one chooses to have, has been perceived as a human right and that anything restricting that was to be opposed.

The one category of audience members that was open to the concept of planning population growth and contraction through the 21st century, was grandparents. Almost to a person when speaking to grandparents, they stated that they were concerned about what the world would be like when their grandchildren became adults. Also, this demographic was obviously beyond having kids, as their decisions were long behind them on the topic.

To lighten the feeling in these rooms, I would do something that both opened up the conversation and also made a very clear and valid point. I told the room that I had a question for the WOMEN ONLY. "Raise your hand if you think there would be fewer children born if women had the primary, even sole, control of how many children to have."

After a few seconds, a small number of hands went up. Then most of the women in the room raised their hands. This happened EVERY TIME. Sometimes it got raucous, sometimes laughter, sometimes arguments with men in the room who pushed back on the idea.

Women bear the responsibility to carry the child and give birth to it. And they typically do most of the parenting which can go on for decades. A key issue going forward is to allow women ever more control over how many children to have. This aligns with a current trend in the developed world and a nascent one in the developing world.

In "Entering the Shift Age" written in 2012, I had a chapter entitled "The Ascendancy of Women". In it I forecast that there would be more growth in gender equality between then and 2025 (13 years) than in the prior 37 years (since 1975). That seems to be occurring in the developed countries. I also wrote:

"There are many dynamics of the Shift Age that clearly point to its being the first age in which women will be truly ascendant. While physical strength may be an advantage men continue to have over women, the old social, cultural and economic power that has traditionally defaulted to men will rapidly fall away…

Most if not all of the social, physical, and cultural forces that have defined men and women since the beginning of the Agricultural Age are in dramatic shift... The Shift Age will be viewed by future historians, male and female, as the first age of humanity when women finally and fully shared in the shaping of history.

The Shift Age will usher in a reconsideration, or an extreme lessening of gender differences and social roles. Perhaps the only differentiation that will be in place is the gender distinction of childbirth. This in fact could actually accentuate and amplify the ascendency of women in a world where many of the global issues we now face flow from the significant population explosion of the last seventy-five years.

I think that the Shift Age will be viewed as the transitional time from the reality of millennia of assumed male dominance to the reality of assumed gender equality and very probably female dominance."

Population Numbers

Here is a chart that shows the ever smaller amount of time between attaining another billion of global population:

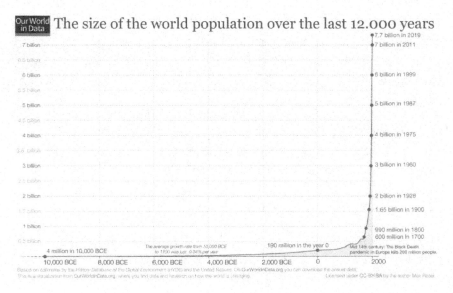

Chart 16.3

It's clear that in the second half of the 20th century the years between billion markers shortened noticeably. The first time interval was 32 years, the next 15 and then 12. As fully discussed in Book 1, the timespan from the 1970s to now is when CO2 PPMs, surface Earth temperatures, and resource depletion all accelerated dramatically. Population growth is a major cause, as the second chart in this chapter points out.

To keep a population at a constant number, the replacement rate requires an average birth rate of 2.1 children per woman. Above this number the total population increases. Below 2.1 it decreases. In many parts of the world birth rates are slowing. The developed countries are mostly under this 2.1 number and countries in Africa and Asia are mostly above.

Project Drawdown, perhaps the most comprehensive study ever conducted on the lowering of GHG emissions, concluded that empowering and educating women is the single most effective way to reduce emissions. Drawdown's number six solution is educating girls and number seven is family planning. Added together they become the number one solution. So it is a major component of moving to a Finite Earth Economy: to empower women and to provide them with what they need to live better and more productive lives.

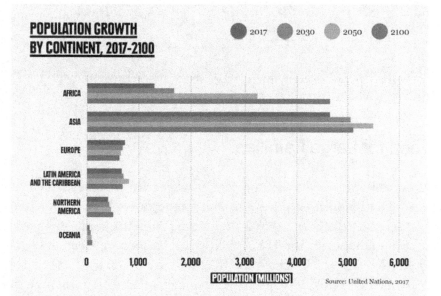

POPULATION GROWTH BY CONTINENT, 2017-2100

2017 2030 2050 2100

AFRICA
ASIA
EUROPE
LATIN AMERICA AND THE CARIBBEAN
NORTHERN AMERICA
OCEANIA

0   1,000   2,000   3,000   4,000   5,000   6,000

POPULATION (MILLIONS)

Source: United Nations, 2017

Chart 16.4

This chart shows that Asia is the most populous continent and Africa is second. Europe will have a lower population in 2100 than today. North and South America will be marginally larger, as will Oceania. The growth in total population to the end of the century will largely be in Africa and Asia. Therefore, if population planning for the next 20 years is considered a significant part of moving to a Finite Earth Economy, most of the work will need to take place on these two continents.

"Increased adoption of reproductive healthcare and family planning is an essential component to achieve the United Nations 2015 medium goal population of 9.7 billion people by 2050. If investment in family planning, particularly in low-income countries, does not materialize, the world's population could come closer to the high projection, adding another 1 billion people to the planet. We model the impact of this solution based on the difference in how much energy, building space, food, waste and transportation would be used in a world with little to no investment to family planning, compared to one in which the projection of 9.7 billion is realized. The resulting emission reductions could be 123 gigatons of carbon dioxide, at an average annual cost of $10.77 per user in low-income countries. Because educating girls has an important impact on the use of family-planning, we allocate 50 percent of the total potential emissions reductions to each solutions — 59.6 gigatons a piece."[6]

Source: Page 79, *Drawdown*, edited by Paul Hawken

Here are some historical ways we have looked at births, children and population.

Need for Large Families

Until well into the 20th century, when modern medicine successfully created anti-biotics, vaccines and many other life-saving treatments, it was necessary to have a good number of kids for some of them to reach adulthood. In the U.S. family size declined between 1800 and 1900 from 7.0 to 3.5 children. In 1900, six to nine of every 1000 women died in childbirth, and one in five children died during the first

5 years of life. Family size increased from 1940 until 1957, when the average number of children per family peaked at 3.7.[7]

FIGURE 1. Fertility rates* — United States, 1917–1997

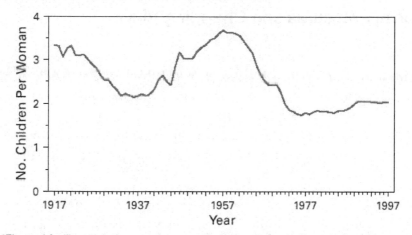

*The total fertility rate is the sum of age-specific birth rates for single years of age for women aged 14–49 years. The birth rates for single years of age used to compute total fertility rates are based on births adjusted for underregistration for all years and on population estimates adjusted for undernumeration; therefore, they cannot be compared with birth rates and fertility rates.

This continues to be true in parts of Africa and Asia where healthcare, clean water, and sanitary conditions are less than optimal. It is these two continents where education and available healthcare can drive down the number of births per woman.

Organized religions encourage large families to create increasing numbers of followers. The history of religion is to expand toward ever more believers. This was to ensure the survival of the religion's dogma and to support its clergy. There is no major traditional religion that embraces birth control. This author finds that to be both short-sighted and lacking in compassion. Large families are a contributor to poverty which results in unnecessary suffering.

This is an area where the metaphor of Earth as a spaceship is especially relevant. Spaceships have finite resources. These resources must sustain the number of crew members onboard. The crew must make do with what is onboard. If the crew had many children, there would be a strain on the spaceship's finite resources. If the crew on the spaceship consumed 170% of onboard resources (as humanity currently does with the Earth's resources) then, at some point, the mission would have to be aborted.

Social Pressures and Changing Views

In the US, even after the curing of most infant and childhood diseases by the last   third of the 20th century, legacy thinking that couples must have children remained the norm. Governments and corporations needed population growth to help sustain GDP growth. Religions needed to keep generating demand for their services. Most adults had grown up with one or more siblings, which therefore set expectations for their own procreation.

In the 1980s there was a social phenomenon called DINK (Double Income, No Kids). With the economy exploding, husbands and ever more wives had the opportunity to work to generate income and wealth. Many couples chose not to have children. These couples had to swim upstream against the perceptions of societal norms. Couples with children often accused the DINK couples of being selfish and indulgent.

Today, with the growing knowledge and acceptance of climate change, couples are choosing to have only one child or none at all. I have spoken to numerous millennial couples who are making the conscious decision to not have kids.

38% of 18- to 29-year-old Americans believe that a couple should consider the risks of climate change before deciding to have kids. "I can't have a child unless I am seriously, seriously convinced that we are on a different path." Other reasons for millennials to put off having children is that they just do everything later than prior generations, and that a major deciding factor to remaining childless is economic.[8]

In all the discussions I have had with a dozen or so couples, the topic of climate change came up. Frequently it was stated that they felt it "selfish" and "self-indulgent" for couples to have more than two children. It was selfish in that it did not take into account the rapidly growing population and the need to move to planetary sustainability.

So, the ethics and morality of having children in the US completely changed in 40 years.

A Finite Earth Economy acknowledges that the earth is just that, finite. Until we can successfully get to a sustainable way of life, bringing children into this world only adds crew members who will be largely dependent until at least the age of 21. As stated in the beginning of this chapter, bringing a child into this world adds some 9,000 tons of CO_2 during the life of the child.[9]

The highest level, longest term way to think of population as it relates to moving to a Finite Earth Economy is to set goals for the rest of the century. Today we are just short of 8 billion people on Earth. Rather than expect the growth of population through time to 10 or 11 billion by 2100, we can move to less than 2.1 children per woman, with the goal to have seven billion or fewer people by 2100. No one alive has to die prematurely, nor do they have to fight for limited resources with billions of new dependent passengers on Spaceship Earth. Moving to a goal of seven billion inhabitants by 2100 will be, along with actually moving to a Finite Earth Economy by 2030, a way to ensure that there will be a vibrant civilization in 2100.

Takeaways

1. Population and population growth must be an active topic of discussion as we move to a Finite Earth Economy.

2. Having a child is the single biggest decision a person can make in their lifetime relative to carbon emissions.

3. Empowering and educating women is a primary area of focus in our move to a Finite Earth Economy

4. Women should have the full say on how many children to have.

5. Africa and Asia represent the biggest challenge (and largest opportunity) for population planning.

6. All of our social mores, beliefs and practices around having children must be examined from our new perspective in the 21st century.

7. Until religions acknowledge that population planning is essential for civilization to continue, they will be in moral opposition to the need to move to a Finite Earth Economy. Overpopulation causes more suffering than population planning.

Chapter Seventeen

Nation States

"We have now entered the global stage of human evolution. This means that we are moving toward more global integration economically, socially, culturally and ultimately, politically. This Flow to Global means that the function and independence of nation-states around the world must and will change... The nation state will not go away, but its independence and separateness from the rest of the world is over." – David Houle, *"Entering the Shift Age"* 2012

Note/Preface: In the Introduction, I stated that I am a futurist, not a policy expert, or a politician, government official, or someone intimately knowledgeable (except as an observer) of the workings of government. As a futurist, I focus on ascendant historical trends, technology, how technology affects humanity, big picture high level concepts and flows. As a futurist on climate change, over the last five years I have written one book "This Spaceship Earth" plus this trilogy, spoken around the world on the topic, and co-founded a climate crisis nonprofit. As a futurist it would be a professional dereliction of duty if I did not address climate change as it is the single biggest factor affecting humanity's future.

What this chapter is about are the major actions that nation states must undertake to confront climate change and move to a Finite Earth Economy. These activities must then be converted into national legislation, policy and tax codes.

The underlying premise, stated in Book 1 and Book 2, is that if humanity wants to enjoy a thriving civilization at the end of this century, it must move as much as possible to a Finite Earth Economy by 2030.

The metrics on which nation states measure themselves were stated in Chapter 8 in Book 2. We also touched on historical examples when a common enemy triggered rapid and large collective actions. The key example is how quickly the U.S. converted to a wartime economy after Pearl Harbor. It's a precedent that proves the ability of governments and citizens to understand that massive, urgent change is called for… and to quickly step up together and make it happen.

In this chapter, and in the next three (Corporations, States and Provinces, Cities), are high level concepts. Each concept suggests actions and policies to implement. Every concept discussed is based on the simple fork in the road in which we find ourselves: oblivion or a chance at a vibrant, evolving civilization by the end of the 21st century… the Earth Century.

Clearly every nation state is different. This means that every country will have different paths to attaining a Finite Earth Economy. Smaller countries will lead the larger countries on implementation simply due to their inherent nimbleness. Some countries are too poor or underdeveloped to take many of the steps suggested. So, the way to read this chapter is to recall the top ten countries by economic output and the top ten GHG emitting countries listed in Chapter 9 of Book 2. It is these countries that this chapter is largely focused on.

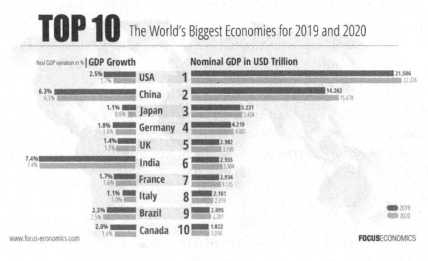

Chart 17.1

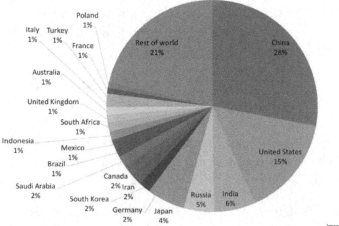

Chart 17.2

High Level Change

Most of the legal, economic and taxation policies of today's nation states are based upon a post WWII view of the world, and the integrated global economy that evolved following the death of the Soviet Union. Those 20th century paradigms must be reinvented to successfully navigate the above-mentioned fork in the road. Massive directional change is required. Bold action is essential.

For a decade it has been clear that we have entered a time when legacy thinking is what is holding back human evolution. I wrote an entire chapter in "Entering the Shift Age" about how 2010-2020 (I named it the *Transformation Decade*), would be an era when legacy thinking slowly collapses across the board, and is replaced by the new realization of 21st Century thought.

"Legacy thinking is viewing the present and future through thoughts from the past. A simple metaphor for this would be the act of rowing a boat. You are looking back to where you have been with your back facing where you are going. You are backing into the future, looking at the past."

"The Transformation Decade of 2010-2020 is when much of what was accepted in the last century will be dramatically altered by the forces of the Shift Age. When I say that this is the first real decade of twenty-first century thought, I mean that 100 years from now, historians will look back at the decade we now live in, 2010-2020, as the decade that began the major story lines of the twenty-first century. Thinking about almost everything will undergo a change in nature, shape, form or character." – "Entering the Shift Age" 2012

We are letting go of the legacy thinking of the past and creating new thinking for the 21st century… the Earth Century.

Carbon Fee and Dividend

As stated in Book 2, this is the first action any nation state should take, starting in 2020 with implementation in 2021. This single action will trigger a rapid decline in fossil fuel use. The key component is to start with a significant amount of tax per ton of carbon emissions.

Here's how it works:

A fee on carbon dioxide emissions is implanted at the first point where fossil fuels enter the economy, meaning the mine, well or port. Exports to countries without comparable carbon pricing systems would receive rebates for carbon taxes paid, while imports from such countries would face fees on the carbon content of their products. Proceeds from such fees will benefit citizens in the form of larger carbon dividends. All the proceeds from this carbon tax would be returned to the people on an annual basis.

In the U.S., we support the immediate implementation of the conservative initiative, the Climate Leadership Council (CLC). It starts at $40 per ton. The Citizen's Climate Lobby (CCL) initiative has a similar plan, but advocates for smaller fees.

The benefit of this Fee and Dividend is that it is a financial redistribution from those that create carbon emissions to the citizens of the country. It has been estimated that the $40/ton fee will generate an annual dividend of approximately $2,000 for each family of four. The dividend offsets the increased cost consumers will pay at the pump and elsewhere. Consumers will support the raising of the fee over time because their dividends will increase accordingly. This fee and dividend scheme will most help the less affluent because the amount is relatively more significant for a family earning $30,000 than for one earning $300,000.

There are other benefits to Fee and Dividend. It places a fee on carbon emissions where it enters the national economy. So it could replace tariffs. For example, if the U.S. imports a shipment of anything from, say China, and China does not have a carbon fee, then that shipment will be taxed based upon the estimated tons of emissions generated to manufacture and transport that shipment of goods. The trigger benefit of this is that countries will move to a Carbon Fee and Dividend to avoid these payments at the borders of other countries.

Carbon Fee and Dividend will incent everyone from multinational corporations to Mom and Pop shops to reduce their burning of fossil fuels. Nation states that are first movers will reap the most immediate benefits, triggering faster implementation globally.

Tax Codes

A key principle is to tax the behaviors we want less of. Tax what is bad. We want to lower carbon emissions, so we base the tax code on that. All entities that pay taxes will pay based upon their GHG emissions (with a primary focus on CO_2) by ton and/or the percent they are carbon neutral. In addition, all currently tax-exempt entities such as religious groups and nonprofits must pay some discounted rate based upon their carbon footprint. (This is a fundamental change in thinking that will occur when we move away from a growth economy based upon profit.)

Nonprofits don't get taxed today because they exist in a profit-based economy. In a Finite Earth Economy, taxes are based on GHG emissions. I can't think of a religious or nonprofit group that doesn't emit GHGs. The "nonprofit" will change to "non-carbon". Non-carbon entities don't pay a carbon tax. The goal is to have this be so successful that, sometime in the 2030s, there will have to be another major change in tax laws once we attain 70% clean energy globally.

Step 1: Measure or estimate a bench line of prior year emissions in terms of total tons of emissions.

Step 2: Set a dollar amount of tax per ton of emissions (here we use $500 per ton). For example, if the average American is responsible for the emission of 16 tons per annum, they pay $8,000. If a large, polluting corporation emits 100,000 tons, they will pay $50,000,000.

Step 3: As technological measurement of emissions becomes more exact; a greater granularity can be injected into the tax code.

Step 4: Carbon tax exemptions for carbon neutral entities (who must prove they are carbon neutral).

Step 5: Tax credits for carbon negative entities.

Steps 4 and 5 will accelerate the development of carbon offsets. Carbon offsets in this context will be investing in anything or any technology that captures CO_2. This could be the purchase of planted trees, or any technology that directly captures CO_2 (either as it is being emitted, or by removing it from the atmosphere). Initially there will be a huge demand for these carbon offsets as people will be playing catch-up for years.

Incentives

1. Tax credits for electric vehicles (EVs), solar installations and any product or service that lowers carbon emissions. These credits will be fully available in the first year of the transition. So an EV purchased in 2021 will have that full credit deducted from taxes in that year. As an example, the average internal combustion engine car emits approximately 4.6 metric tons a year, so the tax credit would be $4,600.

2. A "cash for clunkers" payment of $2,000 for any car on the road that gets less than 35 mpg for highway driving.

3. A set amount in the low thousands of dollars the first year a tax paying entity reaches a carbon neutral footprint.

Subsidies

Most national subsidies are based upon Legacy Thinking. In the US for example, there are subsidies paid to the fossil fuel industry and the industrial agriculture industry.

The most recent annual number for the US is $26 billion paid to the fossil fuel industry. We immediately move 100% away from fossil fuel companies and award them to clean energy providers including nuclear.

The most recent annual number for subsidizing industrial farming in the US is $20 billion. We immediately move 100% to non-industrial regenerative, small and local farms. Meat producers using the industrial model will get $0 government subsidies. Instead they will receive subsidies or tax credits to be used to migrate to a regenerative model for grazing.

In the Population chapter we showed that the decision to have a child is the single most impactful carbon emissions decision a family can make. Given that 2.1 children per child-bearing woman is the replacement rate, there is a $20,000 one-time carbon fee due for child number three and any subsequent child.

National Service

Given the urgency of our climate change situation... we are starting way too late and have so much to do, it is a good time to create a non-military national service. Every country can create some form of national service that confronts climate change issues. In Book 2 we referenced precedents such as the US mobilization for WWII, and the large employment and national service initiatives of the New Deal.

To create the crew consciousness necessary to succeed in the effort to move to a Finite Earth Economy, we propose creating different "Spaceship Earth Crews".

- Conservation Crew
- Clean-up Crew
- Carbon Offset Crew
- Energy Efficiency Crew

Conservation Crew

This Crew is focused on conserving and restoring natural habitats. They could be deployed anywhere and in any way that restores nature. They could help farmers move from land used for mono crops to land that is designated for intercropping. Crews could be deployed to transform large vacant areas in cities to greenfield parks and gardens.

Clean-up Crew

This Crew focuses on the clean-up of plastics, cigarette butts and any waste that has made its way into natural habitats. Countries that have ocean shorelines can clean their beaches, and out to the 12 nautical miles or 13.8 miles to international waters. There will be a hazmat group that restores toxic areas back to natural health.

Carbon Offset Crew

This Crew will plant trees anywhere and everywhere. They will install any new technological carbon capture devices in the most strategic places possible. There are "artificial trees" that can absorb carbon at many times the rate of natural trees. Due to the new GHG emissions-based tax code, creating large carbon offsets will be critically important.

Energy Efficiency Crew

This crew is responsible for retrofitting buildings and installed infrastructure to be more energy efficient. They could start with all national government buildings, then move to the state and local levels of government. Their work in the private sector could be partially paid for by businesses who would therefore get some form of tax rebate or credit for funds spent.

The economic and other benefits of these Crews could solve multiple social and economic problems. We use US metrics here:

1. Everyone gets paid $15 an hour. These mandatory service crews will largely be comprised of people who are 18 to 24 years old. Unemployed older citizens will be offered an employment opportunity if they choose to take it.

2. For every year worked (minimum two years), each crew member would get a $10,000 credit for higher education, to be used within 5 years. This addresses the horribly high costs of college and universities. This benefit would be capped at $50,000 for five years.

3. Widely implemented, this will create a national spirit of collaboration and cooperation… all working toward the common good.

4. A byproduct will be the crew consciousness that is essential for the timely and successful move to a Finite Earth Economy.

5. The crews will supply the work force needed as we make strategic plans for the Strategic Retreat that is immediately ahead in the 2020s.

Different countries will have different agendas. Poorer countries facing sea level rise might prioritize coastline restoration and infrastructure projects to support the inevitable migration of citizens inland. Other countries, whose agricultural production is declining due to climate change, might focus on countrywide substitutions to crops that thrive in warmer weather… whatever is appropriate and relevant for that country.

All of the above are major undertakings. If they are all implemented, it would ensure the move to a Finite Earth Economy by 2030. We should start with these and other big ideas and then move to more granular policies once we learn from the experience of implementation.

For humanity to have a future at the end of this century, for humanity to successfully move to a Finite Earth Economy by 2030, massive effort at the nation state level is absolutely crucial.

Takeaways

1. The move to a Finite Earth Economy necessitates letting go of Legacy Thinking… the mindset that put us into this climate crisis situation. We must create the new and leave the old behind.

2. High level, large scale change is what is called for. Partial efforts, procrastination, half measures and incrementalism will not work.

3. Carbon Fee and Dividend is the first and biggest action nation states should take. ASAP!

4. Recast the tax code based upon carbon emissions not financials.

5. Introduce incentives to help move the private sector toward a lower carbon future.

6. Immediately shift all government subsidies from carbon creating entities to clean and natural ones. Start with ending all fossil fuel subsidies and all industrial farming subsidies.

7. Create non-military national services in all nation states focused on reversing climate change.

8. The magnitude of the climate change reality, and the need to move to a Finite Earth Economy by 2030, necessitates profound changes and political leadership.

Chapter Eighteen

Corporations

"Nation state as a fundamental unit of man's organized life has ceased to be the principal creative force: international banks and multinational corporation are acting and planning in terms that are far in advance of the political concepts of the nation state." – Zbigniew Brezezinski, 1971

In the chapter about Nation States, we emphasized that in this Shift Age, we have entered the global stage of human evolution. That the big problems humanity faces today are all global in scope: climate change, wealth inequality, migration, over population, access to clean air, clean water and food. This is our new stage of evolution. This means that while nation states may always exist, they cannot successfully face these global issues alone.

The only entity that operates above, or inside and outside the nation state is the multinational corporation. The name says it all. They operate in multiple countries. Given that climate change is a planetary, global issue, this suggests that the multinational corporation is the only organized entity (other than the U.N.) that correlates to global issues.

The upside here is that, as governments reach their limits of functionality and authority, corporations can step into the void. They (due to their size, resources and talents), will be leaders in the move to a Finite Earth Economy.

"Saving Civilization!" is not a bad tag line for any multinational corporation.

We start with multinationals, but nation-limited corporations can step up as well. In recent years we have moved from conscious capitalism to actively working for the common good. Enlightened corporations have already moved into this market space and have been well-rewarded by consumers.

In a time of increasingly dysfunctional governments, corporations can swing into action more quickly and efficiently. Given the absolute requirement to move to a Finite Earth Economy, and the ensuing carbon taxes, it serves corporations to get out in front of the global effort. In 2022 and for decades after, the climate crisis will force humanity to take escalating actions to deal with it. Why not take the lead and earn customer loyalty and respect for action taken?

The single most difficult thing for all corporations, large and small, will be to relinquish Legacy Thinking. Legacy Thinking argues for the status quo, for the way things have been done, for the old order of things. The metrics of the past (shareholder value and quarterly profits) are assumed to be relevant in the future. They will not be. They are a major contributor to global warming and the climate crisis.

Given a high degree of inevitability, either we move to a Finite Earth Economy or corporations lose their customer base. Why not embrace it? Why not leverage all that high price talent to carve out a leadership role and competitive advantage? Not greenwashing, but real transformative leadership, with real actions backing up mission statements.

What Corporations Will Need to Do

Corporations currently do four things that they will have to fundamentally alter in the move to a Finite Earth Economy:

1. They are often large GHG emitters. Collectively, they are the largest emitters. Their emissions must be cut drastically and eliminated altogether within a couple of decades. They can lead in this effort and then sell their expertise either through products developed or via consulting services.

2. The raison d'etre of corporations in growth economies is to grow. Grow sales and profits, both of which are financially measured and result in the perception of "success" or "value". In the short term this change will be the hardest to translate. The move from monetary growth measurement to emissions reduction measurement will be the most difficult to undertake.

3. Corporations are designed to sell products. In a finite world, infinite growth is impossible. The realization must be that between now and 2030, no growth in sales is the new target. Again, a difficult cultural and mind set transition.

4. Corporations use marketing and advertising to promote consumption. They will now need to promote non-consumption.

Emissions

Corporations must make the measurement of their GHG emissions, CO_2 in particular, their top priority. To measure is to know. Stating the problem and framing the magnitude of it is the first step. As mentioned in the technology chapter, the ability to collect data from the surface of Earth and from satellites in space currently exists. We now have the technological tools to measure emissions down to single factories and small geographical areas. Rapid deployment will occur when corporations are taxed and fined based on their emissions.

Multinationals will first use metrics to determine in which countries they have the highest emissions. Then they will interact with the national, state and local government authorities to find ways to reduce or offset their emissions.

Any size corporation will then move to overall measurements of emissions and waste. They will identify their levels, locate them, quantify them and work to reduce them.

Note to those that think 'big bad corporations don't care' about climate change and their responsibility in helping to create it: this is why we must move to a carbon emissions/carbon footprint tax base. Corporations work hard and spend significant resources to lower their current taxes which are based on income and profits. Taxing emissions will cause them to expend as much or more on lowering those emissions.

Non-Growth

Profits will be addressed by taxes on emissions. If corporations do not cut emissions, their profits will suffer. Stockholders will pressure corporations to lower emissions because share prices are affected by earnings per share.

In a world where reducing emissions is the top priority, doing so will be good corporate governance. We will be moving from a focus on shareholders to a focus on stakeholders. Successful corporations will be the ones that focus on stakeholders by valuing the well-being of people and nature.

Sales

Sales as a profession will continue, but it will undergo a major transformation. In the business-to-business world where salespeople play a central role, touting a product's ability to cut costs and improve productivity will no longer be enough. Because the customer will have to pay taxes on emissions, they will be looking for low or zero emission products. Their cost of ownership of a product will increase significantly if a product emits GHGs. So corporations will invest in the research and development of non-emitting products. Resulting in a flood of innovations.

Selling with the new metrics of the Finite Earth Economy creates an entirely new sense of collaboration between seller and buyer, between company and customers.

Advertising and Marketing

Advertising exists to persuade people to buy things… more and more products and services. Our consumer society is a major driver of GHG emissions and global warming. In a Finite Earth Economy, the only things we want more of are clean energy, and products and services that decrease GHG emissions.

As now constituted, advertising and marketing are in direct conflict with saving civilization. By design, they make the situation worse not better. They must evolve to creating messaging about strides in cutting emissions, and about innovations that deliver products and services with zero carbon footprints. An educated populace will make buying decisions based on the actual amount of carbon emitted in a product's manufacture. Brand loyalty will depend on the percentage the company has lowered carbon emissions over the prior year.

Buying used is better than buying new in a Finite Earth Economy. Companies and industries that create business models based on the resale of items will benefit with products that have no carbon production footprint.

In a Finite Earth Economy there will be plenty of opportunities for organizations that produce tools to measure emissions and waste… and can quantify the composition of those emission and waste streams at a granular level. Consumer Reports magazine has been successful for decades by providing consumers with quality and price comparisons. Those rankings will grow to include analyses of emissions and waste in production and in use. Being an informed buyer around these issues will become the norm as people become crew members, and understand how critical these stats are to civilization's survival. J.D. Power is a company that rates quality and customer satisfaction. A company that gets a J.D. Power acknowledgement of being the best product in a category, immediately advertises that achievement.

In a Finite Earth Economy, governments will measure the carbon footprint and emissions of every company and base their tax rates on those numbers. This data can then be used to inform citizens which companies are in fact reducing emissions by migrating to clean energy sources.

Unfortunately, the past decades have shown the hypocrisy of companies that advertise their 'greenness'. This is called greenwashing and it is the type of thing (undocumented and misleading claims) that has always plagued advertising.

Given a widespread ignorance of the dynamics of emissions and sustainable product design, companies have taken advantage with green marketing campaigns that are nothing but window dressing. It is annoying to see fossil fuel companies pitching their environmental efforts. "Hey, look at this initiative we are funding! We are paying people in white lab coats to find a better way to convert algae into energy." These are small investments done for the PR value. Meanwhile the companies are investing billions to extract oil and gas from deep in the ocean or by dredging tar sands. This type of disingenuous and even fraudulent advertising will no longer be tolerated.

The natural gas industry has launched campaigns declaring their product 'clean' energy. It is not! Natural gas generates CO_2 emissions when it is burned. Gas leaked during extraction and transport emits methane, a much more powerful GHG than CO_2. It is not clean. It is only less dirty than petroleum or coal. Doing less bad is not the same as doing good.

In a Finite Earth Economy it will be law that the only way a company can promote itself as green, or concerned with the environment, is if it has achieved zero emissions, or it can verify that it has a carbon neutral footprint. Advertisers could say they have, for example, "lowered our emissions last year by X% or by X tons", but they cannot say they are environmental in any way unless they are carbon neutral.

Finite Earth Economy non-consumer groups will publish and promote the actual tonnage of GHG emissions of every public company and as many private companies as possible. Facing a climate catastrophe occurring because of GHG emissions, companies that are found to misrepresent emissions or greenness will be both fined and punished in the marketplace. Just as many informed consumers do not buy products from companies that use child labor, customers will not buy from companies that do not cut emissions.

If this is not enough, every ad will be required by law to show the total tonnage of GHGs that the advertiser company emitted the prior year. And the amount that it declined over the last two years. They will not be allowed to do this in fine print. Following the example of warnings on cigarette packages, these emission numbers will be prominently visible. This will drive corporations to improve in the current year as they will be encumbered by their prior year's emissions.

The Transition

At least for the next five years, we cannot fine corporations for emissions as it will take time to fully set up carbon offsets, and change the context and metrics around operating in a Finite Earth Economy versus a Growth Economy. Corporations will be paying dearly with taxes on emissions unless they lower them. Customers will want to buy from companies that are measurably working toward the goal of 70% clean energy by 2030.

The companies that will really take off are the ones – most likely in the technology sector – that create products or services that capture, drawdown or cleanse atmospheric carbon. Corporations that create technologies that contribute significantly to saving civilization will be lavishly rewarded.

In a Finite Earth Economy, successful corporations:

- will openly and aggressively move to reduce their carbon emissions.

- will market that fact relative to their competitors.

- will create good will and market share by establishing leadership positions.

- will reinvent advertising and marketing in a world of ever lower consumption.

Takeaways

1. Corporations must lead the way to a Finite Earth Economy. They can move more quickly than governments.

2. Corporations account for the greatest amount of GHG of any sector in society so their emission reduction efforts are the most important.

3. Corporations will adjust their operations and business models to conform to the new metrics of a Finite Earth Economy.

4. Sales, Advertising and Marketing will all change from consumption and financial metrics to carbon metrics.

5. Corporations that can develop and market the purchase of used rather than new products will be successful.

6. The transition will be difficult.

Chapter Nineteen

States and Provinces

"If your profits are based on consumption, where's your incentive to reduce electricity use?" — Rick Kriseman

States and Provinces are the largest organized governmental entities below the nation state. There are many nation state laws and tax policies that govern citizens. The states and provinces comply and align with "The law of the land". Then states and provinces build on and augment those laws and policies.

Sometimes states lead a nation in policy implementation. A U.S. example is the state of California. States and provinces will focus on and emphasize different policies due to regional variations. In this chapter we look at the major contributions states and provinces can make to the move to a Finite Earth Economy by 2030.

Different states will make different contributions depending on the size of their populations, topography, economic and industrial makeup, shorelines, rivers, number of vehicles... many aspects including, unfortunately, their politics. The author lives in Florida, the state at the highest risk for climate change impact. This state will therefore be discussed in detail. The Age of Climate Change will affect Florida differently than Arizona, New Hampshire or Wisconsin, for example.

What follows is a list of high level actions that every state can take to move to a Finite Earth Economy by 2030.

Lead by Example

Make commitments to decrease carbon emissions and set an aggressive timetable to accomplish them. Examples might be that all state buildings will be 50% clean energy by 2027 and 70% by 2030.

Commit to the entire state vehicle fleet transitioning to electric or hydrogen fuel cell by 2030.

Commit to decreasing all streams of non-recyclable waste by 50% from the benchmark year of 2020. This can be incentivized via fees/taxes on households with overages.

Make it clear to all state employees and citizens that we have entered a new age – The Age of Climate Change. And we must

all evolve our mindsets from being passive passengers to active crew members on Spaceship Earth.

The legacy thinking, systems and business models that exist either must adapt to a non-carbon paradigm or they will fail, because the rules are changing in the move to a Finite Earth Economy.

No one can serve in state government if they take election funding from any entity that supports the continuance of a fossil fuel economy. Every politician can be forgiven for his or her past, but from the effective date of the law onward, they cannot take those donations and hold office.

Legacy thinking must be replaced. What worked then, does not work now. The in place businesses, utilities, land use policies and transportation models are what created the unfolding carbon crisis. Why hold on to the sources of the problem?

Transportation

Levy a purchase tax on all internal combustion engine vehicles. Based not on cost, but upon miles per gallon which convert to carbon emissions. A small compact that gets 45 miles per gallon is taxed much lower than a big pick-up truck that gets 15 miles per gallon. An imported hybrid that costs $50,000 will get taxed less than a $25,000 gas-powered SUV. The revenues from this tax can be used to build state-wide charging stations for electric vehicles (EVs).

Highway Lane Use

Inform state residents that in the near future (establish a date), highway lanes will be allocated to encourage EVs. For example, a province could declare that in five years, the left lane will be for EVs with two or more passengers only; the second most left lane will be for EVs and hybrids only, and the right lane(s) will be for gas-powered vehicles.

Relative to national interstate trucking, there will be mileage assessments for gas and diesel trucks when entering the state, similar to today's weighing stations.

Taxes

Some states and provinces have a personal income tax. These states would move from an income tax to a carbon footprint tax. The tax would be similar in form as the national tax discussed in Chapter 18, but would be an add-on (just like the current income taxes). Aside from all the reasons detailed in this book about the necessity to move to a Finite Earth Economy, states could use this tax to verify the cleanliness of their air. We already measure pollution and pollen levels by location on a daily basis. Lower emissions, cleaner air, healthier people.

Trees

States and provinces could provide trees to plant at a nominal cost. These would be the types of trees that survive best in the state's climate and also absorb the greatest amount of CO2.

Residents could, for example, receive a $5 tax credit for every $10 paid to the state for the purchase and planting of the trees.

Energy

States must support any and all sources of clean energy. There is a critical issue with the current grid system in the US. The utilities that provide power via fossil fuels have been doing so for decades and have developed working relationships with politicians and governments.

It is inevitable that clean, distributed energy will be running alongside the in place centralized grid model. Current utility business models are based upon 20th century industrialization and centralization. They are designated utility monopolies. These business models will only exist in 2030 if they are at least 70% clean energy.

Distributed energy is more resilient to natural disasters. States should encourage small scale, distributed solar, wind and Generation 4 nuclear, along with larger solar and wind farms, and nuclear reactors. Today's grid providers will adapt, change or fail without the support of the public, the investment community and non-purchased politicians.

The author lives in Florida, the "sunshine state" and the third most populous state in the U.S. It's logical to assume solar energy would be quite prevalent. Yet Florida ranks 19th in solar energy deployed, behind states with much less including New York and Massachusetts. This is due purely to the fact that the electric utilities have an installed model that is highly profitable and have resisted changes to it.

In a New York Times article about Florida's energy situation, these realities are clearly stated. Below are some quotes from the article with a link to it.

Headline: "Florida's Utilities Keep Homeowners from Making the Most of Solar Power" byline - Ivan Penn

"Florida calls itself the Sunshine State. But when it comes to the use of solar power, it trails 19 states, including not-so-sunny Massachusetts, New Jersey, New York and Maryland.

Solar experts and environmentalists blame the state's utilities.

The utilities have hindered potential rivals seeking to offer residential solar power. They have spent tens of millions of dollars on lobbying, ad campaigns and political contributions. And when homeowners purchase solar equipment, the utilities have delayed connecting the systems for months...

Solar energy is widely considered an essential part of addressing climate change by weaning the electric grid from fossil fuels. California, a clean energy trendsetter, last year became the first state to require solar power for all new homes.

In Florida, utilities make money on virtually all aspects of the electricity system — producing the power, transmitting it, selling it and delivering it. And critics say the companies have much at stake in preserving that control.

"I've had electric utility executives say with a straight face that we can't have solar power in Florida because we have so many cloudy days," said Representative Kathy Castor, a Democrat from the Tampa area. "I have watched as other states have surpassed us. I think that is largely because of the political influence of the investor-owned utilities…

Mayor Rick Kriseman of St. Petersburg — the site of Duke Energy's Florida headquarters — has argued for changing the way utilities are regulated so they would embrace more energy efficiency, residential solar power and energy storage. The companies essentially see the solar-equipped homeowner as a competitor, not a customer, he said.

From 2014 through the end of May, Florida's four largest investor-owned utilities together spent more than $57 million on campaign contributions, according to an analysis by Integrity Florida, a nonprofit research organization, and the Energy and Policy Institute, a watchdog group. FPL, the state's largest utility, accounted for $31 million of that total…

solar energy accounted for only 1 percent of electricity generation in Florida last year, far less than the 19 percent in California and nearly 11 percent in Vermont and Massachusetts, the association said. The state relies largely on natural gas, and several utilities get as much as a quarter of their power from coal…

"I would say that none of Florida's utilities are enthusiastic about their customers' deploying solar," said Stephen Smith, executive director of the Southern Alliance for Clean Energy. "I am not surprised at the horror stories."

Scott McIntyre, chief executive of Solar Energy Management, a statewide leader in commercial solar power based in St. Petersburg, said the gains the state appeared to be making were little more than a facade.

"Florida is not going to do any type of energy policy that benefits consumers, not for a long time," Mr. McIntyre said. "They just keep making the hurdles higher and higher."[1]   https://www.nytimes.com/2019/07/07/business/energy-environment/florida-solar-power.html

There will be a repeal of existing laws or regulations limiting the installation of solar energy panels on rooftops. It will be illegal for any city, neighborhood or homeowner's association to prohibit the installation of solar panels. Instead there will be guidelines as to types of installations, no prohibitions.

There will be a state carbon tax which means that all taxpayers will want to reduce their carbon footprints. Part of any carbon footprint is the source of the electricity consumed. Residents will not want to pay tax to the state government because their utility burns fossil fuels. The utility companies that do move to all clean energy, or at least 70% by 2030, will be aligned with the tax policies of the state.

Water

States can provide a tax credit for the installation of "grey water" systems in homes and businesses. Grey water is water that has been used more than once. Fresh water is supplied to a home or business. That's what flows out when you turn on the tap. That water can then be recycled within the system (from the kitchen and bathroom sinks and the showers) to be used to flush toilets or water gardens and plants. Currently there are zoning laws in place in many areas of the world that do not allow grey water systems, established when water was perceived as unlimited and close to free.[2]

Allow citizens and businesses to capture rainwater. In urban areas there is storm runoff due to impermeable surfaces of sidewalks, driveways, parking lots and roads. In nature, much more of the water is absorbed into the ground. Captured rain water will not be lost in 'run-off', and can be used for plants, landscaping and gardens. In rural areas, allowing the capture of rain water for flushing toilets and irrigating gardens will help those using well water at a time when the water table is falling.

Land Use

This is tied closely to water. In real estate developments all around the U.S. there are laws saying that homeowners MUST have green lawns. This makes no sense relative to water conservation. Watering the average lawn uses as much water as filling 16 swimming pools!

Green lawns were imported from the United Kingdom, where it rains often, and the soil is rich. There isn't much need for sprinkler systems there. It does not make sense to have lawns in desert cities like Los Angeles or Phoenix. Anyone flying into these cities will see hundreds of miles of brown, tan and yellow desert. Then, suddenly, the landscape turns green with lawns. Purely a result of unnatural amounts of water being used for an outmoded legacy concept.

Perhaps lawn could be allowed IF they are irrigated using a grey water system.

Building Permits, Rules and Regulations

The energy source and carbon footprint for all buildings must be transparent to city and state authorities (as well as the property owners). Expected carbon footprint per annum for the life of a mortgage must be fully disclosed as part of any real estate transaction. Property that is not energy efficient will be much less desirable, and that will be reflected in lower market prices. This will prompt developers and flippers to focus some of their investments toward energy conservation.

Agriculture

In farm states and provinces, policies will be established to motivate the move to sustainable farming practices. Below market loans, tax abatements, and whatever else that will accelerate the rapid conversion from industrial farming and monocrops (that typically spray nitrogen-based fertilizers and toxic weed killers) to regenerative agriculture, permaculture and intercropping (that naturally maintain and promote nutrient rich soil and pest control).

This chapter is a high-level list of some of the more urgent and essential matters that must be addressed for the move to a Finite Earth Economy by 2030. There will be variations between states, but there's a common need to reward actions that create a rapid move away from fossil fuels.

The key in any state or province is the alignment of government officials with the greater body politic to lower carbon emissions ASAP!

Takeaways

1. States and Provinces will play a critical part in the move to a Finite Earth Economy.

2. States and Provinces have oversight over many aspects of business, utilities, land use and more… all of which must evolve.

3. There will be significant variations from state to state and province to province.

4. Transportation, energy, land use, water use, zoning, taxation and approval of all building projects will comprise initial essentials.

5. The laws, taxation codes and removal of politicians receiving fossil fuel campaign contributions will create a crucial alignment between the government and the body politic.

Chapter Twenty

Cities

"But a city is more than a place in space, it is a drama in time." — Patrick Geddes

"City farming is not only possible, it is the very definition of the kind of meaningful, sustainable innovation we will need to meet the grand challenges of the 21st century: climate change; population growth; ageing population; urbanization; rising demand for energy, food and water; poverty; and access to healthcare." — Frans van Houten

"The city needs a car like a fish needs a bicycle." — Dean Kamen

Cities are the future of humanity. It was in 2013 that, for the first time, more than 50% of all humanity lived in an urban setting. That number is now 54%. Almost every population forecast increases that to 70% by sometime in the 2040s.

What makes this even more dynamic is the simultaneous population growth. The 2019 number of city dwellers is 54% of 7.8 billion, which is 3.9 billion. In the 2040s it will be 70% of 9 billion or 6.3 billion. This means that there will be 2.4 billion more inhabitants of cities between now and 2045. This translates into 96 million people moving to or being born into urban environments EVERY YEAR between 2020 and 2045!

Megacities today represent a greater percentage of global GDP and global waste than the per capita worldwide. A megacity is defined as having 10 million or more inhabitants.[1]

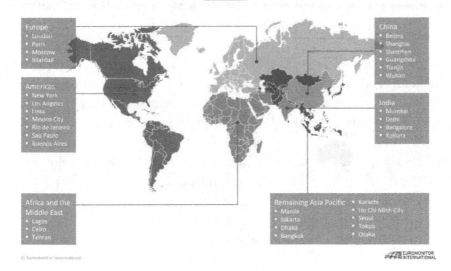

Current megacities:
breakdown by countries/regions

Chart 20.1 Euromonitor[2]

In 2010, the 27 existing megacities represented 6.7% of the total human population, but generated 14.6% of global GDP. They generated 12.6% of all waste. In 1990 there were 10 megacities. Today there are 33. It is expected that there will be 43 in 2030 and close to 60 by 2050. So, as we ramp up to move to a Finite Earth Economy by 2030, cities in general and megacities specifically, are at the epicenter of all the changes that need to occur.[3]

In Chapter 15, Strategic Retreat, we showed that seven of the largest 20 cities in the world by population are at high risk of Sea Level Rise (SLR). We must, therefore, draw the conclusion that sea level rise will dramatically disrupt economies in the not too distant future. So strategic retreat is not just about moving people.

Every city vulnerable to SLR must begin planning for climate migration. Those cities at high risk of SLR must start to make long range plans to address the situation, initially with defensive strategies to buy time. If we can move to a Finite Earth Economy by 2030, many cities may not have to relocate.

Cities have a split personality when it comes to GHG emissions. To use New York City as an example, today 8.3 million people live in the city, and another 12 million inhabit the total market area including the suburbs in New York, New Jersey and Connecticut. As this chart shows, the per capita of CO_2 is much lower than the national average for people living in the city proper. This is due to a number of reasons, including the availability of public transportation, more compact living spaces, less external exposure to outside elements (think a 25 story high rise with 10 units per floor versus a single family home), and a much lower than national average level of car ownership. The amount of CO_2 per household in the suburbs is four times greater. Suburban residents must have cars to get around (even if it is just to the commuter rail). They are heating and cooling (on average) many more square feet of space than city apartment dwellers. And single family homes are exposed to the elements on all four side plus their roofs.

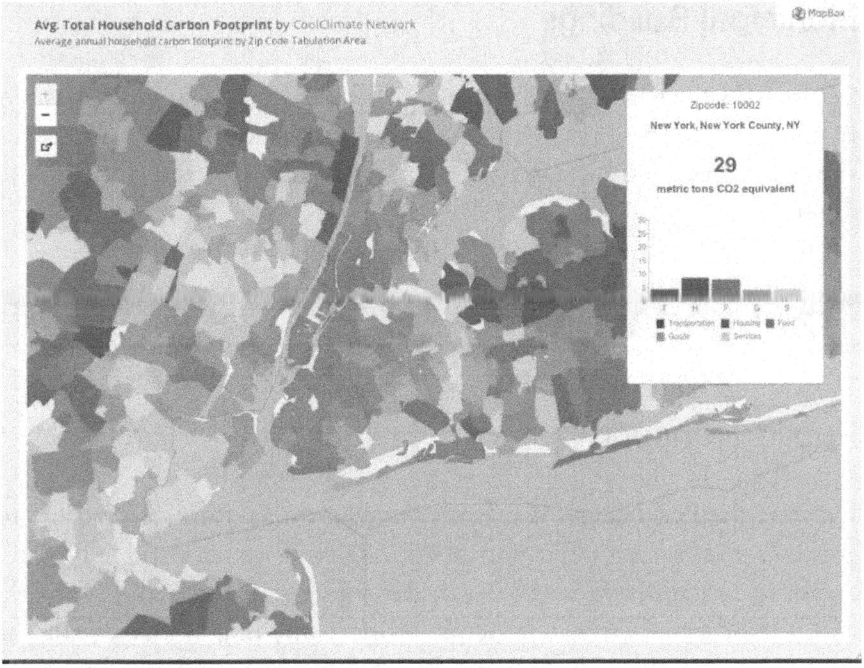

Avg. Total Household Carbon Footprint by CoolClimate Network
Average annual household carbon footprint by Zip Code Tabulation Area

MapBox

Zipcode: 10002
New York, New York County, NY

29

metric tons CO2 equivalent

Chart 20.2

"Metropolitan areas look like carbon footprint hurricanes, with dark green, low-carbon urban cores surrounded by red, high-carbon suburbs," explained researcher Christopher Jones. The carbon footprints of suburban homes tend to be about double the national average; those of cities, only half. The average household in Manhattan, for example, contributes 32 metric tons of carbon emissions annually, while nearby Great Neck averages 72.5 metric tons per home.[4]

Action Plan for Cities

There are a number of steps cities can take to move to a Finite Earth Economy by 2030. Here is a list, far from exhaustive, of basic actions to take targeted to the reduction of emissions.

Municipal Buildings

Convert ALL city owned structures to clean energy as soon as possible. Install green roofs, solar panels, wind turbines and new industrial strength direct carbon capture technologies on as many rooftops as possible.

Transportation

Public Transport

Increase the percentage of the city population that has easy access to good, fast, reliable public transport. This will increase ridership and decrease emissions. Here's a list of the top 10 cities around the world with the best mass transit systems. Notice only one US city made the list: Chicago. NYC came in at number 11. The criteria used to rank the cities included efficiency (how quickly you can move from Point A to Point B), affordability, safety, cleanliness and size (number of passengers per year). These mass transit systems not only include trains and buses, but also boats, trams, bike lanes and cable cars.[5]

1. Singapore
2. London
3. Hong Kong
4. Paris
5. Madrid
6. Chicago
7. Tokyo
8. Dubai
9. Shanghai
10. Zurich

Electric Vehicles

Replace all city vehicles with plug-in electric (can be phased in over time). Then, via public relations campaigns, influence residents to either live a car-free life, or set up incentives for residents who must have a car to either rent for short periods of time or purchase an EV. This can be done by a variety of actions:

- add a city purchase tax for any gas-powered vehicle.

- in public and private garages, mandate both free charging stations right next to handicap parking places and a double tier pricing policy with the cost of parking an EV significantly below a gas-powered vehicle.

- zone strategic locations for the rental of bicycles, electric scooters and EVs.

Zoning and Land Use

Eliminate any zoning restrictions on installation of solar, particularly in the high-emissions suburbs.

Eliminate any restrictions on grey water systems.

Eliminate any zoning restriction on reuse. An example would be the conversion of empty big box stores into vertical indoor gardens. This would allow some percentage of the city to buy food that avoids emissions from transport. It also would create a local relationship between growers and purchasers, and bring as much agricultural production as possible back into the city. This is important because 60% and then 70% of humanity will live in cities. That will put an incredible strain on supply chains and increase CO2 emissions for food transport.

Reuse could also mean the conversion of such outdated things as large suburban McMansions into multi-unit housing, or closed shopping malls into educational institutions. Basically, retrofitting real estate from its 20th century intended use (when emissions were not a consideration) to today when emissions and their measurements are primary concerns.

Institute any and all carbon capture technologies, applicable to each and every neighborhood. These include planting trees, painting roof tops and external walls white with paint pigmented with titanium dioxide, and installing large carbon capture technologies on the ground and on the roofs of buildings. NOTE: I first wrote about titanium dioxide in a blog post 12 years ago after reading about it in a New York Times article. Titanium dioxide is not an expensive chemical. It's used to pigment paint into a brilliant opaque white. What was discovered 12 years ago was that it also has the surprising capability of "eating" air pollution up to a distance of eight feet from the paint. So titanium dioxide not only reflects sunlight helping to cool buildings, it also cleans the nearby atmosphere of pollutants.

Set a limit on waste per household and/or per person per annum. This should be a fairly easy metric to enforce as the city usually has oversight on the pick-up of garbage. Any waste over this amount must be paid for with a waste tax. If some people are caught dumping garbage so it won't be counted, they will be subject to significant fines.

Support Neighborhood Efforts

All cities are comprised of neighborhoods. City governments should offer financial support, materials and in-person expertise to any neighborhood wanting to establish gardens, or compost collection stations, or shared reuse centers (clothes, furniture etc.).

This will be covered in much more detail in the next chapter titled "Personal – Becoming Crew".

Taxation

There are income taxes at the local and state level. For cities that have an income tax, align it with the carbon model of the country and the state. Cities can choose to have a penalty tax for excessive emissions, excessive waste and the use of private internal combustion engine vehicles. An annual tax for any vehicle that gets less than 30 miles per gallon.

New Urban Developments

In a Finite Earth Economy, new developments will be designed from the ground up to be carbon free. The capability to do this already exists. A brilliant example of this is Babcock Ranch in Florida. We visited this amazing development a year and a half ago when it was selling its first homes.

Simply put, Babcock Ranch is not just a development, it is a complete town that is 100% carbon free. All the energy comes from a 340,000 solar panel (soon to be doubled in size) farm that will power the entire town to its expected final population of 50,000. All homes are energy efficient, from their architecture to their heating and cooling systems and landscaping. Each garage has an EV charging station. The entire town has free gigabyte per second broadband WIFI.

Babcock has established a transportation plan that includes solar powered "free taxis" and an autonomous EV bus that is driverless and transports up to 10 people. Organic gardens grow produce for the restaurants and stores in the downtown part of the development.

Babcock Ranch shows what can be done when starting from scratch to build a carbon free living environment. New carbon-free developments and retrofitted communities will become more popular as the new taxes on emissions make them more affordable than communities that continue to emit carbon.[6]

In summary, cities will continue to grow larger and exert more influence over all aspects of life, culture and economics. They must lead in the conversion to a Finite Earth Economy.

Takeaways

1. Cities are where the majority of humanity lives. They also can move more quickly than states and nations, so they must lead the other levels of government to a Finite Earth Economy.

2. Every year an increasing percentage of humans will live in cities.

3. Megacities – those with more than 10 million people – will grow in number, and will increasingly be where human culture and commerce occur.

4. There are a number of ways that city governance can trigger change and they all must be considered. What they do and how they do it will vary by location, climate and size of the city.

Chapter Twenty-One

Personal (Becoming Crew)

"There are no passengers on spaceship earth. We are all crew." – Marshall
McLuhan

"Saving civilization is not a spectator sport. We all have to get involved." –
Lester Brown

As stated previously, in order to successfully deploy the Finite Earth Economy, we will need to make significant changes in how we live. That will require top down changes (technology, regulations, tax policies, investments, infrastructure) and bottom up changes (personal habits, evolved status markers, local zoning laws, increased civic participation).

Bottom up is just as important as top down… not only for the effects of widespread adoption of reduce, reuse, recycle; but for the byproducts. In this instance, actions precede belief system changes. Once the process is begun, people understand what it means to be a crew member on Spaceship Earth. They see opportunities everywhere to improve processes and reduce their carbon footprints. And they become active in their communities… activists who learn to use the tools available to them to influence those in a top down position. This is the journey this chapter takes: from personal to neighborhood to community to city hall and beyond.

With any luck, a virtuous cycle ensues. Top down to bottom up and back again. Each feeding the other. Every sequence decreasing emissions and reducing waste, until crew consciousness, a stewardship mindset, pervades every strata of society.[1]

The two most important concepts to keep in mind are crew consciousness and a game plan to quickly reduce your carbon footprint and GHG emissions. Aside from doing the right things, in a Finite Earth Economy you are going to be taxed on emissions. Combine the two concepts: crew members en masse quickly changing how they live and reducing their carbon footprints. That synergy will be significant in its effectiveness in reaching our goals.

Let's start with some items where we have data.

Individual Action

One person can make a difference by thinking and acting as an active crew member. Start small and simple. As these tasks become habit, they no longer feel strange. They happen automatically. Then you can add more to your daily activities.

"We need to evolve from being passive passengers on Cruise Ship Earth to being active crew members on Spaceship Earth." – Tim Rumage, Planetary Ethicist

"If you think that you are too small to make a difference, try spending the night with a mosquito in the room." – Dalai Lama

Becoming a Crew Member on Spaceship Earth

The fundamental cause of climate change is siloed thinking. This means that humans who live in Growth Economies do not have a cause and effect connection with nature and Earth. We are disconnected from the most complex, elegant and interconnected systems that comprise the environment in which we live.

In the neighborhood where I live in Florida, the garbage gets picked up on Tuesdays. This means that I put out the garbage cans and the recycling bins every Tuesday morning and by the evening the waste is gone. In reality it isn't gone, it has just been moved to a place that is out of my sight… where the majority of it will either be incinerated or buried in a landfill. But it is still on Spaceship Earth. Out of sight, out of mind.

Think about sailing on the cruise ship referenced in the quote above. We sit down at all you can eat buffets without knowing where the food has come from. We have no idea what happens to the waste from those meals. It's a great metaphor for how we live. Where did the food come from that you ate today? What will happen to what is not eaten and any other waste from the meal? You likely don't know. You are an unconscious consumer… a part of the process that is causing global warming, our climate crisis and the sixth extinction of species. Why? We are not active participants in the process. All we do is purchase, consume and throw away. That is a consumer process, not how Spaceship Earth was designed to work.

In a growth economy, our role is to consume. Because you are thought of and treated as a consumer, you are targeted by hundreds of advertisements every day. We are persuaded to buy. We are taught that buying makes us feel better about ourselves. That buying ever more somehow makes us cooler, smarter, more desirable. We are even sold the idea that "shopping" is a productive activity.

The truth is that consumption for consumption's sake is not cool. The subsequent GHG emissions and waste streams are the cause of global warming and the climate emergency we are experiencing. Crew consciousness replaces this disconnected, self-centered, consumer mindset, with one that honors and protects the earth, and preserves it for our progeny.

Becoming crew has two major outcomes:

1. You become aware of the consequences of your actions. You no longer can say you didn't realize. You take responsibility.

2. As the number of crew members increases, the benefits that accrue to the biosphere also increases accordingly. Scale kicks in. This aspect of being crew makes it clear that individual acts, when scaled, do make an important difference.

Here are a couple of examples that can directly be connected to CO_2 emissions.

Think about how much beef you consume. As an American I was raised in the land of the cheeseburger. How many hundreds, if not thousands of these have I consumed in my life?

When I started to study climate change, I realized I was not acting as crew. Once I became crew, I learned that every quarter pound hamburger I ate generated six pounds of CO_2. This is largely due to today's industrialized farming and ranching. My favorite meal was a great cheeseburger with crispy fries. (Note: the cheese and fries add more CO_2, but we will only address the emissions from the beef.)

I ate at least two quarter to half-pound cheeseburgers a week. So, I consumed approximately 12 ounces of beef twice a week. In a year (52 weeks), I consumed roughly 40 pounds of beef. Multiply that by 24 pounds of CO_2 per pound of beef, and that emits 960 pounds of CO_2 into the atmosphere. Almost a half a ton of CO_2! Just because I like cheeseburgers. Now I average no more than two per month which translates into nine pounds of beef or 216 pounds of CO_2... a reduction of 744 pounds of CO2 emitted into the atmosphere So that is acting as a crew member.

Now, think how crew consciousness scaled changes the equation. What if 10 million people did exactly what I did? That's 10 million times 744 pounds. Which translates into 7.44 billion pounds or 3.72 million tons less CO2 into the atmosphere by just one crew action! One crew action taken by just 3% of the U.S. population. What if 10 or 20% agreed to reduce beef consumption by 75%?

Crew consciousness makes one feel responsible for Spaceship Earth, the only home we have. Acting as a crew member, and recruiting others to become crew, creates a group sense of aligned action.

Paying Your Electricity Bill

Think about how you pay your electricity bill. You approach the monthly paying of the bill with a pre-conception of the price. Say you have been paying between $90 and $110 a month. That is the price range you are conditioned to pay. If the bill is in that price range, you pay it without thinking about it. Now, let's imagine you get a bill for $130. Your first reaction is to think about why it might be higher. Maybe it was colder or hotter than normal, or maybe you had guests in the house using the hot water, etc. You remember that and pay the bill.

How much energy are you paying for? When I ask people this question, less than 5% have any idea.

A crew member might take the 2% challenge. What this means is that instead of looking at the dollar amount, you look at the amount of energy your household consumed. In the average $100 per month home, it's around 1,000 kWh of electricity. Commit to consuming 2% less the next month, or 980 kWh. Then keep lowering usage by 2% a month for a total of six months – you are now consuming 880 kWh per month. This is easy to do once you have adopted crew consciousness. You won't experience any discomfort. Turn off lights in empty rooms. Raise or lower the thermostat by a couple of degrees depending on the season. Unplug devices when not in use.

Every eight households that do this saves one entire household's worth of energy. The average household wastes 35% of energy used. For U.S. households, approximately 75% of energy used in the home comes from fossil fuels.[2]

This is when being crew scales up and answers "Does what I do make a difference?" Just getting rid of one third of the waste reduces usage by 12%. One million crew households will save 12 million kWhs every month, which translates roughly into 7,000 barrels of oil or 900 tons of CO_2.

Plastic

In the U.S., and in many other parts of the world, 90% of plastic is NOT recycled. Every household has more than enough plastic in the home already. Instead of buying anything packaged in plastic, pack your lunch in a reused plastic container. Bring one to the restaurant for your left-overs. Carry several to a store that sells in bulk. Eliminating one pound of new plastic removes three pounds of CO_2 emissions.

The average US household throws away 185 pounds of plastic per year which emits 555 pounds of CO_2 via the manufacturing and transport processes. Reuse as much plastic as possible and get others to do so as well. If just 10 million households cut plastic consumption by 25%, there will be a net saving of 230,000 tons of CO_2 emissions.

Cutting beef consumption, decreasing electricity used, and repurposing plastic are just three examples of how crew members can make a real difference. Another way of looking at this is that it saves money to lower CO_2 emissions. All the above actions save money as well as CO_2. Plus, in a Finite Earth Economy, CO_2 emissions will be taxed.

In the **Addendum** are several links and resources that can provide guidance as to where to cut CO_2 emissions in your life.

Consumer Consciousness

"Human beings are good at many things, but thinking about our species as a whole is not one of our strong points." – Sir David Attenborough

The choices we make as consumers can make a huge difference. We need to have a paradigm shift regarding how we spend our money. Think of every dollar you spend as a vote. A product purchased is you fulfilling a need for yourself or your family… and in another space, it's a data point on a product sales report. This data is used to decide what a company will make more or less of, and what will be on the shelves in the stores. Next time you are at the store consider this concept. We, as consumers, need to act as crew, making informed decisions, voting with every dollar, making our demands for earth conscious products known.

Visit your local farmer's market once a week. Bring your own reusable cloth bags. Pick up fresh fruits and veggies without putting anything in a plastic bag or clamshell. Less waste coming in means less waste going out.

Ditch plastic baggies. Replace them with reused jelly jars, saved yogurt tubs and other plastic containers.

Stop using paper towels and napkins. Buy some cloth napkins for the dinner table, and cut up old bed sheets and towels to use around the kitchen.

If you live close enough, bike to work. If you don't, start a carpool or take mass transit. Investigate whether your employer will allow you to work from home a few days per week. Buy a hybrid or electric vehicle.

If you eat eggs, raise chickens. Keep bees. They will provide you with honey. The chickens will eat any ticks and fleas in your yard, and the bees will pollinate your flower and vegetable gardens.

Many power utilities are now offering the chance for their customers to enroll in a "Green Power" program where you pay a little extra (about $60 a year) for your power, and the utility has to prove to you and the government at the end of the year that it used power from alternative energy sources (wind, solar, hydro, etc.). If they can't document their alternative energy use, they are fined. Therefore you are holding them accountable to invest in alternatives.

You decide who you buy from. Every decision you make from the car you drive, to the toys you give your children, to the clothes you wear has a ripple effect on Spaceship Earth.

Every dollar we spend is a vote:

- We are voting for where our food is produced – if it must be transported long distances, or is locally and organically grown.

- We are voting for the ingredients that comprise each product we buy – are they toxic chemicals or non-biodegradable plastics?

- We are voting for how much waste is generated in the manufacture of each product.

- We are voting for the entire impact the product will have throughout its useful life and after it's disposed of.

We need to educate ourselves about the foods and products we consume… about all the "stuff" that we surround ourselves with. Go into your bathroom, open every cabinet and drawer, and pull out all the items. Place them on the floor and categorize them: lotions, shampoo and conditioner, makeup, razors, shaving cream, hair bands, contact solution, clippers, travel bottles, meds, nail polish, etc. Step back and look at all of the small plastic containers. Think back to how many of those came packaged in a box or sealed in plastic. Decide which products you can do without. Promise yourself not to buy any more of them. Then decide which items are necessary and research options to buy that product in bulk. Bulk buys use less packaging and require fewer purchases (and fewer trips to the store) per year.

Once you are done with this highly achievable project, make a plan and set aside time to do this to other rooms in your home. How about your kitchen, closet or garage? You can simplify your life by paring down to your lifestyle's necessities.

Think before you buy. Do you really need a new sweater? How many are sitting in your dresser drawer? Is your urge for a new sweater prompted by a need to stay warm, or is it because your sweater is no longer in style? "Fast fashion" is a major contributor to global warming. The fashion industry was responsible for 1.7 billion tons of carbon dioxide in 2015. Buy quality over quantity. Have two sweaters in that dresser drawer instead of six.[3]

As we move to a Finite Earth Economy, wearing the latest fashion will lose its cachet. Respect will be earned by those who purchase smart… those who look beyond themselves and consider the consequences of their purchases on Spaceship Earth and all of its inhabitants.

Neighborhood

"Do what you can, with what you have, where you are." – Theodore Roosevelt

Make a concerted effort to be part of your neighborhood community. Get to know your neighbors, if you don't already. Start a neighborhood environmental team. Begin by socializing… picnics and barbecues (you can grill vegetables). Then grow by presenting educational seminars. There are many environmental groups with chapters in every city. The Citizens Climate Lobby (CCL), the Climate Reality Project, and the Pachamama Alliance are among many who will happily present to your neighborhood group.

Host native plant giveaways. Get creative. In the process, you will plant seeds of awareness, and make new friends and allies. Then you're ready to tackle some slightly more ambitious projects like these:

Alternative Transportation Challenge

Use friendly competition to get neighbors biking, walking and using public transit more often… good for the environment and your health.

Bike and Transit Buddy Programs

Match experienced cyclists (Mentors) with people who would like to bike more, but could use some guidance (New Riders). Match experienced public transit riders with interested, but less experienced riders. Matching programs decrease car usage and CO2 emissions by increasing bike riding and public transportation use.

Bulk Energy Purchasing

Save money and reduce your carbon emissions by participating in a community purchasing program. These programs give homeowners volume discounts when their neighborhoods decide to go solar.

Bulk Food Buying Clubs

Bulk food purchasing provides healthy options at a price cheaper than many grocery stores. In addition, it builds community as people work together to select and distribute food. Reducing trips to the grocery store decreases the use of fossil fuels and emissions from personal vehicles.

Clean Up and Recycling Event

Provide the neighborhood an opportunity to dispose of items that are not collected curbside. Cleanup events minimize waste by matching hard-to-recycle items with willing recyclers. Neighborhood clean-up events:

- Reduce the amount of waste hauled by trucks or freight to landfills.

- Reduce the amount of toxic waste (think cigarette butts) that can flow into the natural environment and cause harm to other species.

- Save energy by recycling hard to extract materials.

Community Composting

Install a large composting bin in a central place where neighbors can drop off their food scraps. Over time the food scraps decompose into nutrient-rich soil that can be used in the community garden. This reduces the amount of food waste which generates methane in our landfills.

NOTE: Composting has many advantages. It keeps food scraps out of landfills. The aerobic process of composting does not produce methane because methane-producing microbes are not active in the presence of oxygen. The nutrient rich soil produced by composting sequesters $CO2$, and it precludes the need to apply nitrogen fertilizers which tend to run off the soil and befoul our waterways.[4]

Community Vegetable Gardens

Some people have more garden space than they can manage, and do not have an interest in gardening. Other people may want to garden, but don't have the space. Matching up these two groups increases local vegetable production and reduces transportation emissions.

Neighborhood Green Business Agreements

Create a green buying club. Approach local merchants and promise to give them all your business if they promise to take actions to reduce their carbon footprints and/or stock sustainable products in bulk. Think about how many half gallon plastic containers of laundry detergent you buy every year. Now cut that number down to one because you can take it in for a refill at the supermarket.

Neighborhood Tool & Resource Sharing Programs

Provide lending libraries of various tools and resources for neighbors to borrow. These programs connect neighbors with the tools they need without the cost of buying them. Tool and Resource sharing:

- Reduces consumption which cuts down on multiple emissions processes such as manufacturing of products and transporting them.

Neighborhood Tree Planting

Trees provide an invaluable resource since they naturally sequester carbon. Shade provided by trees reduces the need for cooling and thereby decreases energy use in the summer. Create work projects for kids to learn early that planting a tree is always a good thing, and cutting one down is not.

Weatherization Teams

Form a group of neighbors to work together to weatherize each other's homes. This may include sealing air leaks or installing vinyl storm windows. Identify homes that need weatherizing. Check in with your senior neighbors or others who may need extra help around the home.

Local Institutions

Church, schools, city hall… reach out to them to establish programs. Get to know the people who run these institutions. Find out which city administrators are open to making changes that reduce GHG gases and waste. Here are a few suggested programs you can initiate:

- Replace plastic cutlery in schools, church cafeterias, etc. with reusable flat ware.

- Establish community gardens for congregations, at senior citizen centers, at schools, in neighborhood parks, etc.

- Lobby the mayor's office for zoning changes that allow gray water to be used to irrigate gardens, solar panels to be installed just about anywhere, mini-wind turbines on buildings, more bike lanes, etc.

- Put on educational events at churches and schools. Partner with environmental groups. Teach kids and adults what you've learned to do.

Get involved in your community. Educate yourself on the positions of local politicians, and vote! Make it clear that addressing climate change and moving to a finite earth economy is the most important criteria on who you vote for.

Again, carbon emissions as a base for taxation will have a profound effect on all these institutions and groups. They will want to lower their own emissions by working more closely with those that have the expertise. A perfect opportunity to spread crew consciousness.

Policy

"It isn't enough to do our best, sometimes we have to do what is required." – Winston Churchill

Write your congress person, lobby for changes in the laws, learn about the voting records of elected officials and VOTE!

Democracy is based on an educated electorate, therefore, to have a true democracy we need to ensure climate change education. For a deep dive on the importance of education for the new world we all are living in, see this "reimagination" of the entire educational system.

Voting is the cornerstone of democracy. However, these days in America, only about 60% of the eligible population votes during presidential election years, and about 40% vote during midterm elections. For this reason, we have elected officials who are expanding the climate crisis instead of solving it... and we are partially to blame. Our collective voices are what will make change for our communities. For some, this will mean running for a government position within their county, city or state. For most, it means taking a few personal steps like those outlined above. Either way is a win for humanity.

Takeaways

1. Addressing climate change by moving to a Finite Earth Economy must occur on all levels: global down to personal.

2. There are a variety of things to do as an individual, as a member of a family as a resident of a neighborhood, and as a citizen of a town.

3. Carbon emissions are the culprit. Cut them as much as possible, as soon as possible.

4. Start to think of yourself as a crew member of Spaceship Earth. There is no resupply ship coming, so we must make do with what we have on board.

5. As a crew member, consider the cause and effect of all that you do: what you buy, what you eat, how you transport yourself and your family.

6. Stand tall in the knowledge that your crew actions make a difference.

7. Recruit other crew members. Collectively you will boost each other and become ever more crew conscious.

8. Crew actions will quickly become habits requiring little thought or extra effort.

9. As you interact with other crew members, new ways of thinking will arise to replace your old consumer mind set.

10. This Spaceship Earth is your responsibility. Crew with vigor at all levels of life, from personal habits to national and global politics.

Chapter Twenty-Two

Urgency! Urgency! Urgency

"We are called to be architects of the future, not its victims." – R. Buckminster Fuller

"How did you go bankrupt? Two ways. Gradually, then suddenly." – Ernest Hemingway,
The Sun Also Rises

We have been experiencing climate change gradually these past 15 to twenty years or so. We didn't see it or feel it before then, although it measurably started some 50 years ago. Many of us have been concerned for decades that massive climate change was inevitable. Many others of us have not really been paying attention, or discounted the danger, or even doubted the reality of it.

Today it has become undeniable (to most of us). Our future, and the futures of our children and grandchildren, is uncertain at best. We must come together and face this situation collectively. We must rise to the challenge of becoming crew members on Spaceship Earth.

The metaphor of Earth as a spaceship is a good one for us all to keep in mind. There's no resupply ship coming. We have to live within the finite resources of the spaceship. The choice is before us. We can remain passive passengers on Spaceship Earth; which means we continue to live as we always have, unconcerned about the consequences of our actions. Or we step up and take the responsibility of becoming active crew members. And make changes to how we live... constantly aware of the choices we make, and the GHG emissions and waste streams that result.

To stay with the spaceship metaphor a bit longer, sirens are blaring, and red lights are flashing:

- Hundreds of heat records are broken every year.

- 150 species go extinct every day.

- Coral reefs are dying off due to ocean warming and acidification.

- 96% of humanity breathes fossil fuel polluted air, and 10,000 people a day die as a consequence.

- Major cities are running out of fresh water.

- "500 year storms" happen every few years.

Those who are crew respond to the emergency and take actions to stabilize the ship so that we all can continue to live in a civilization worthy of the name. Those who are passengers dismiss the lights and sirens as alarmist, and take the position that there is nothing they can do that will make a difference.

We know that some line has been crossed, some centuries old equilibrium has been thrown out of alignment.

To restate some science from Book 1, if ALL carbon emissions were ended the day you are reading this, the earth would continue to warm for another two to three decades. That is because 600 gigatons of resident CO_2 have accumulated in the atmosphere above the pre-Industrial Age level. There is a true urgency to dramatically lower all emissions ASAP, and to develop and deploy natural and technological cleansing and sequestration systems.

We must do both as quickly as possible. We don't know what climate tipping points have been crossed, only that some already have. We don't know how many more we are on course to cross. Once crossed there is no turning back. As shown in the charts in Book 1, we have been living beyond the Earth's capacity to regenerate for 40 years. We have put ourselves into a planetary hole. We are behind in our reaction by decades.

When we started to write this book, the goal was to move past all the discussion about whether there is climate change. To move past the "save the planet" anthropomorphic naivete. To answer the questions "What can we do?" "What can I do?" The goal was to move to the solution side, at all levels of human organization from the global down to the local. As a futurist, I spend my life looking at trends, doing pattern recognition, connecting dots where others see no connection, looking at what is in ascendency in the moment, in the now, to see what these ascendant flows and forces will be both short term (one to five years) and long term (five to 15 years), and to do so with accuracy.

Having actively spent five years in research, writing, speaking and thinking about climate change, it became clear to me that, as a futurist, I had to do a deep analysis because climate is the most significant influence over humanity's future. As a futurist, a goal of mine has always been to not just forecast what will happen, but to suggest ways to deal with it. In the case of climate, it is so all-encompassing, that meant examining all levels of society (individuals, families, businesses and governments) and determining what they will need to do to avoid disaster. That was the impetus for this trilogy.

However, during the six months it took to research and write this book, we became alarmed, that things are much worse than we thought. We have had a number of conversations about whether it was too late to avoid the climate crisis, that we are already in a runaway hothouse scenario. We agreed that the only position we could take was to not give up hope until we've done all that we can. To do all that we can to help humanity avoid the end of civilization as we know it.

We started this book with great urgency. We wanted to get it out to readers in three months, as soon as possible. But as we dug deeper, and unearthed ever more alarming information, we decided that meeting self-imposed deadlines was less important than getting it right.

One thing we learned was that there is not much hard data on carbon metrics except at the global and national levels. The degree of granularity… at the personal level, at the household level, at the corporate level just isn't available because it isn't being recorded. Because emissions have not been considered important enough to track at that scale.

The research on GHG emissions that we found was general, approximate and not related in measured ways with economics. This is to be expected for two reasons. First, until recently when it has undeniably become a clear and present danger, climate change was still being debated. That debate has largely been concluded now (with one notable, unfortunate exception – President Trump). The second reason is an entrenched information infrastructure designed to measure the things that matter in a growth economy – cost, profit margins, productivity, return on investment, revenue, etc.

As we discussed in Book 2, we are measuring the wrong things. The Finite Earth Economy requires exact measurement of emissions at the same level of detail, and as comprehensively, as we currently track financial transactions. Again and again, when we researched "How much CO2 is emitted when X happens?" or "How much CO2 is emitted per unit of X?" we either came up with no answer or a variety of vague ones. We found phrases like "not doing X is the equivalent to driving Y miles on the highway." In what type of vehicle? A 15 mpg truck or a 50 mpg hybrid? Didn't say.

A clear example that brought this home to both of us was when we were trying to ascertain how much CO2 gets emitted for the six-ounce fish filet you order in a restaurant. There is some consistency that a quarter pound hamburger triggers six pounds of CO2 as there has been intense measurement of industrial farming. So, we decided to see the comparison with fish. We found several estimates, that we approximated into four pounds of CO2. The number feels soft. Why is there no hard data? Because the business that owns the boat that caught the fish that became that six-ounce filet measures salaries, and gasoline costs, and boat maintenance, etc. They know exactly what their cost is per pound of fish delivered to market. Those are the metrics of a Growth Economy. CO2 is not measured, it is approximated, if accounted for at all.

So here we are, needing to move to a Finite Earth Economy that is structured to lower carbon, decrease waste, and cleanse the atmosphere of CO2… and we are just at the beginning of developing accurate metrics to do so.

In Book 2, we emphasized that metrics are key and that there are all kinds of technologies coming online to enable quantification of emissions. We have until 2030 to move as much as we can to a Finite Earth Economy. It is imperative that we establish exactly how much is being emitted at all levels of the economy by 2022 so we can make intelligent choices from then on.

Another factor that has created urgency in this effort is what is going on in the world relative to "the fight against climate change". First, we must stop using words like fight, beat, defeat as they are irrelevant in facing climate change. Humanity cannot fight climate change; we can face it and move as rapidly as possible to "save ourselves from ourselves". Military language is exactly what is not needed. We are not going to defeat climate change. All we can do is change how we live on this planet as quickly as possible to save civilization, and to slow and ultimately reverse climate change over the next century or two.

This book is being written as 20+ Democrats are running for President of the United States; some of whom are making climate a major plank in their platforms. And as the Sunrise Coalition and the Extinction Rebellion are gaining traction. And Greta Thunberg is on a non-stop journey to trigger increasingly effective demonstrations aimed at the powers that be to change policy. Climate scientists continue to sound the alarm. Environmentalists are saying, "See? We told you!" That's all good. And now it is time to define and outline the path forward.

A victory lap is permissible, but we don't have time for more than one. We have a gigantic problem to solve.

We have all heard the phrase, "the first step is identifying that there is a problem."

That's been done. Climate change is real and threatens the existence of thousands of species including humanity. Now we need to solve the problem.

The purpose of this book is to recognize that our Growth Economies are the root cause of climate change. Growth at all costs, powered by fossil fuels, and the deliberate provocation of ever more consumption, have put us in this do or die situation.

We must move with urgency to the solution: a Finite Earth Economy. We have framed the need to move from the causal Growth Economy based upon capital, to a Finite Earth Economy based upon carbon reduction.

When the tax code is based upon carbon emissions, when the economy is based upon the essential goal of ever lower consumption, then the market will adapt. Measurement, capture, cleansing and drawdown technologies will quickly be developed and deployed when the free market focuses on reducing emissions.

The market will jump in with new business models, ideas, and a collective mission to move to 70% clean energy by 2030. The government, with its new carbon emissions taxes, will quickly alter how people, businesses and governments act.

When an ever-increasing number of people embrace crew consciousness and take responsibility for lowering carbon emissions, a growing sense of manifest destiny should develop. As Paul Hawken asks, "Is it game over or game on?" The climate crisis is an opportunity for us all to step up and become our best selves.

Urgency must become our new normal. Time is short. We have identified three paths forward for humanity. The only viable path forward is to replace the root cause of global warming and our climate crisis. We must replace our growth economies with finite earth economies by 2030.

In the past few years we have come to understand that we have only until 2030 to fundamentally change how we live, if we want civilization as we know it to exist in 2100. This is now the official position of the IPCC of the United Nations, and a growing number of climate change scientists, and certainly our position after the experience of researching and writing this trilogy.

As the Hemingway quote at the top of the chapter suggests, humanity has been experiencing a gradual change in the climate. Now that change is about to become sudden. The suddenness of a climate crisis is upon us. The well-being of all life on Spaceship Earth is at risk.

To quote another great author:

"How did it get so late so soon?

It's night before it's afternoon.

December is here before it's June.

My goodness how the time has flewn.

How did it get so late so soon?"

Dr. Suess

Urgency is required given the situation we are in.

Urgency! Urgency! Urgency!

Chapter Twenty-Three

Saving Civilization

"I've never been at a climate conference where people say, 'that happened slower than I thought it would'." – Christina Hulbe, Climatologist

We live in a transformative time.

The next 20 years, from 2020 to 2040 will be as historically important in human history as the Industrial Revolution, the Enlightenment or the Renaissance. We stand on the threshold of our next evolutionary leap.

Darwin suggested in his 1859 book "The Origin of Species" that in the 5,000 years prior, evolution had accelerated. This was due to the creation of civilization, religion, laws, government, culture, language, education and, of course, commerce. Evolutionary biologists have argued that from 1859 to the early part of the 21st century, in 150 years, humanity has achieved a leap of evolution rivalling the prior 5,000 years.

Today with the unfolding of big data, machine learning, global connectivity and accelerating breakthroughs in medicine, energy and Brain Machine Interface (BMI), we might well achieve another evolutionary leap by the year 2050 or 2060. We are ever accelerating, ever growing in complexity and functionality.

We now stand at the height of human accomplishment. In the last 200 years, extreme poverty has dropped from 90% of humanity to 10%. The level of education has tripled in developing countries and doubled in developed countries. The amount of wealth created since 1900 is much larger than all that went before. Life expectancy has tripled in developing countries and doubled in developed. The list is lengthy. We are at the apex of our journey as homo sapiens on planet Earth.

Unfortunately, as we stand on this evolutionary threshold, we also are facing a global climate crisis. Every prior apex species has collapsed. Most for the same reasons… as byproducts of their success. They died out from overpopulation and resource depletion.[1]

The 6th Extinction Event is well underway. It is of our own making. 200 years ago, there were one billion people on earth. Today we are almost eight times as many. Yet the Earth and its finite resources has not expanded. Living in growth economies, with ever more people, has now run its course, and is the cause of what we must now face, together.

To ensure that our species has a chance to take the next evolutionary step, we must face this climate crisis with urgency. To ensure that those currently under 40 have an opportunity to experience this, we must do an urgent about face and move to a Finite Earth Economy. To keep our grandchildren from having to live in a world where people will fight and kill for water and food, we need to act with urgency.

Now is the time to move from Growth Economies to Finite Earth Economies.

Loosely speaking, Capitalism is now degrading Democracy. It is hard to think of an existing Democracy that has not been corrupted by money. Wealth inequality is putting both Capitalism and Democracy at risk. The endless pursuit of growth has hit the wall of a finite planet. Infinite growth has brought us to this true fork in the road.

"In several decades, humanity will be at a fork in the road: utopia or oblivion. Either a utopia of abundance or an oblivion of destruction." – R. Buckminster Fuller, 1969

It is truly time to let go of the status quo. Business as usual is killing us. We need to create a new form of capitalism that will put humanity, all living things and earth first. This is a huge shift in mind set, and how we conduct our daily lives... and it must be undertaken with incredible speed. It truly is the largest undertaking humanity has ever faced.

We must begin NOW!

"The secret of change is to focus all of your energy, not on fighting the old, but on building the new." – Socrates

Urgency is the state we must move to. The cost of not rising to this challenge is an uninhabitable home – for us, our progeny, and most other life on the planet. We know that approximately 150 species a day are going extinct. We know some 10,000 people a day die from fossil fuel triggered air pollution. We know that the sea level is rising at an accelerating rate. We know that there is an ever increasing financial cost. We know that every day we delay in moving toward a Finite Earth Economy will result in an additional $1 billion of global costs. There is no precedent, so the math isn't certain, but I set this down as a conservative figure. It might turn out to be $10 billion a day, because we have waited so long to take action.

We focus on the financial catastrophe ahead as it will be those with the most money to lose, from moving away from Growth Economies, who will hold on the longest and fight back the hardest.

The World Bank estimates a global $520 billion annual loss in GDP. That's $5 trillion in 10 years, $10 trillion in 20. Direct costs to health could be as high as $4 billion per year by 2030.[2]

Add on top of that the uninsured losses of a few hundred billion dollars a year (insurance and reinsurance companies have already either withdrawn from coverage or are increasing premiums significantly) due to property, crop and infrastructure damage. In other words, if we do nothing Growth Economies will collapse under the stresses. If we step up to the challenge now, we can avoid at least some of these losses. If we do not move to Finite Earth Economies by 2030, we face a downward spiral of increasing costs, declining growth, and massive loss of capital and property.

We must accept the reality that global warming is accelerating faster than expected.

We can no longer use data from five years ago because it is now irrelevant. Five years ago an acceptable target for becoming fossil fuel free was 2050. We now know that is too late.

Instead of despair, let's reach for hope. Instead of resigning ourselves to a fate worse than death, let's use this as an opportunity to take that next evolutionary leap forward. This climate crisis is a golden opportunity to become the best we can be… to overhaul how we live, to not only fix the climate, but to fix many of society's ills.

Once we set our new direction, changes on the plus side will accelerate. People will become active crew members on Spaceship Earth… working together. Cooperation and collaboration instead of competition. Corporations will continue to compete – they will compete against each other in how quickly and completely they can lower their GHG emissions. 1,000,000 trees will be planted every day. Politicians not fully on board will be quickly voted out of office.

Entrepreneurs will develop new business models, new technologies to capture carbon, and alternative ways to operate in a carbon-free environment. Investment capital will pour in as a sense of manifest destiny takes hold.

This global effort, led and coordinated by the 20 largest economies, will drastically reduce conflicts around the world as humanity works collectively on common goals. The collective effort to move to Finite Earth Economies will generate a new mindset of cooperation.

Basically, in the all-out effort to "save ourselves from ourselves", we unleash a species turnaround that dwarfs the Industrial Revolution, the Enlightenment and the Renaissance.

"All of our differences are being overshadowed by the reality that we are all together on Spaceship Earth." – *This Spaceship Earth,* **Houle and Rumage**

What if we valued community more than competition? What if we stopped measuring our self-worth by the number of possessions we have? What if from the top to the bottom of societies across the world, we decided to stop living in a materialistic wasteland, and instead decided to become more conscious of how we produce, consume, work, relate and live? To achieve that transition, we must change our consciousness. We must harness our collective energies of compassion, ingenuity and wisdom to transcend our selfish desires and co-create with others an inclusive, healthy and caring society.

Utopia or Oblivion

As mentioned above Buckminster Fuller first advanced this concept in the late 1960s. He posited that in a few decades we would arrive at a crossroads, and would need to make a choice. We now stand at that crossroads and the choice must be made. It won't be easy, but the potential rewards are enticing.

Let's consider what a post-consumer lifestyle might look like:

There would be an economic boom like we haven't seen since World War II… but with no major wars. A large segment of the world's population would be working on either mitigating climate problems or adapting to them. For the first time in history, we now have access to every other human via smart phones, social media and automatic translating software. Innovations would spread like wildfire.

Just about everything must be redesigned. We'd still consume – but mindfully. Products would be designed from the ground up to be used over and over… repaired when broken or worn. When they must be replaced, the component parts are designed to be fed back into the production cycle, not discarded in a land fill.

The whole energy production and delivery system is overhauled. The centralized grid has been partially replaced by distributed solar and wind generation – near to where the energy is being used. There are no single points of failure. We would have a surplus of clean, free energy!

In the U.S., a large part of the Defense budget is reallocated to the development of technologies to address climate change. The Pentagon determined that this expenditure was more effective in increasing safety and stability than the purchase of more advanced weaponry.

Of course, not everything is rosy. There are refugees – mostly due to flooding, droughts and the resulting famines. Refugees are accepted as an unavoidable byproduct of a warming planet. Climate change has driven home the fact that everything is interconnected (we are all in this together). We change our minds (or "consciousness") and are willing to evaluate our actions before we take them. We ask, "Will this action cause harm to myself or others, or the living systems that sustain us all?"

The greenhouse effects have left us with unpredictable weather patterns, less arable land, and rising seas… so we make do. We rely on home gardens for a significant portion of our diets. Obesity is no longer an issue.

Life is quieter. Electric engines run silently. There are no rush hour traffic jams. Aided by technology, most people can work from wherever they wish. There are driverless electric taxis and many fewer cars on the roads. Mass transit has increased dramatically: light rail, electric buses and vans. Bicycle lanes and walking trails have proliferated (and they're being used). Stress has been greatly reduced – no rush hours, more time spent working in and enjoying nature, we walk more and bike more. All of this results in better overall health and a more closely knit social fabric.

There have been reductions in lifestyle – lower carbon footprints, fewer miles traveled, less meat on our plates, etc. The most significant change, though, has been a willingness to pitch in together and seriously face this existential problem. The sharing economy has taken off. People share living spaces, tools, child and elder care… you name it. Neighbors know each other, even in the cities. Depression and suicide trends are reversed.

Much More than Merely 'Surviving'

Most people recognize that our situation is serious, and many realize it's a choice, not a foreordained destiny. If we have the means to choose our future, why are we on the path to a disastrous one? We have choices that could lead not only to survival, but to truly joyful lives for all. Living in abundance during this age of climate change is the new path for humanity.

We must move to a Finite Earth Economy by 2030. We have it in our collective capability to do so. We face our evolutionary destiny now. We can mobilize and move forward so that history books recount that at this time in the first third of the 21st century, humanity stepped up, created the opportunity for civilization to continue, and accomplished the single greatest achievement of our species' history.

Addendum
Charts, Data and Quotes

Charts

3.1 Slide from Paul Hawkens' Drawdown presentation
https://www.drawdown.org/

3.2 Sam Carana, Global Carbon Project
https://www.globalcarbonproject.org/

3.3 Global Footprint Network https://www.overshootday.org/

3.4 https://www.climate.gov/news-features/understanding-climate/climate-change-atmospheric-carbon-dioxide

3.5 https://www.climate.gov/news-features/understanding-climate/climate-change-global-temperature

4.1 https://www.worldbank.org/en/news/feature/2017/12/15/year-in-review-2017-in-12-charts

4.2 Per capita consumer levels Source: Global Footprint Network

4.3 The Circularity Gap Report, Circle Economy, January 2019

5.1 Gross World Product, Our World in Data, Roser 2016
https://ourworldindata.org/economic-growth

5.2 GDP Per Capita, Our World in Data, Roser 2016
https://ourworldindata.org/economic-growth

5.3 World Population Growth, Our World in Data, Roser 2016
https://ourworldindata.org/world-population-growth

5.4 U.S. Consumption and Emissions, Nature Communications 2015
https://www.nature.com/articles/ncomms8714

6.1 and 6.2 2019 Circular Gap Report by the Circular Economy

9.1 Top 10 World Economies https://www.focus-economics.com/blog/the-largest-economies-in-the-world

9.2 Relative Size of National GDPs https://www.weforum.org/agenda/2018/10/the-80-trillion-world-economy-in-one-chart/

9.3 Top 20 Polluting Nations https://www.ucsusa.org/global-warming/science-and-impacts/science/each-countrys-share-of-co2.html

10.1 Top 10 Global Risks https://www.weforum.org/agenda/2019/01/these-are-the-biggest-risks-facing-our-world-in-2019/

10.2 Global Carbon Emission 2018 https://www.theguardian.com/environment/2018/dec/05/brutal-news-global-carbon-emissions-jump-to-all-time-high-in-2018

10.3 Share of Global CO2 Emissions https://www.ucsusa.org/global-warming/science-and-impacts/science/each-countrys-share-of-co2.html

10.4 Cumulative CO2 Emissions https://www.globalcarbonproject.org/carbonbudget/18/files/GCP_CarbonBudget_2018.pdf

10.5 Per Capita CO2 Emissions https://www.statista.com/chart/16292/per-capita-co2-emissions-of-the-largest-economies/

10.6 GHG Emissions by Economic Sector https://www.epa.gov/sites/production/files/2016-05/global_emissions_sector_2015.png

10.7 Resident CO2 in Atmosphere https://www.climate.gov/news-features/understanding-climate/climate-change-atmospheric-carbon-dioxide

10.8 Record CO2 PPM in Atmosphere https://scripps.ucsd.edu/news/carbon-dioxide-levels-hit-record-peak-may

11.1 Global Electricity Generation by Source
https://www.technologyreview.com/magazine/2019/05/

11.2 Renewable Energy Snapshot 2016 to 2017
http://www.ren21.net/wp-content/uploads/2018/06/17-8652_GSR2018_FullReport_web_-1.pdf

11.3 Geothermal Direct Use by Country http://www.ren21.net/wp-content/uploads/2018/06/17-8652_GSR2018_FullReport_web_-1.pdf

11.4 Space-based Solar Power Image
https://earthsky.org/earth/space-based-solar-energy-power-getting-closer-to-reality

11.5 Size of Tesla Gigafactory
https://www.reddit.com/r/teslamotors/comments/2fxw9a/exactly_how_huge_will_the_tesla_gigafactory_be_oc/

11.6 Photo Gigafactory in Reno, NV
https://bgr.com/2016/08/08/tesla-gigafactory-size-cost-elon-musk/

11.7 Tesla Giant Battery in Australia
https://electrek.co/2018/05/11/tesla-giant-battery-australia-reduced-grid-service-cost/

15.1 Top 20 Cities at Risk for Sea Level Rise
https://forms2.rms.com/rs/729-DJX-565/images/fl_cities_coastal_flooding.pdf

16.1 CO2 Footprint per Additional Person
https://www.theguardian.com/environment/2017/jul/12/want-to-fight-climate-change-have-fewer-children

16.2 Global CO2 Emissions 1750 to 2015
https://ourworldindata.org/co2-and-other-greenhouse-gas-emissions

16.3 World Population Last 2,000 Years
https://ourworldindata.org/world-population-growth

16.4 Population Growth by Continent
https://ourworldindata.org/world-population-growth

16.5 – Fertility Rates United States 1917 to 1997
https://www.cdc.gov/nchs/data/statab/natfinal2003.annvol101.pdf

17.1 World's Largest Economies https://www.focus-economics.com/blog/the-largest-economies-in-the-world

17.2 Top 20 Polluting Nations https://www.ucsusa.org/global-warming/science-and-impacts/science/each-countrys-share-of-co2.html

20.1 Megacities https://www.euromonitor.com/global-overview-of-megacities/report

20.2 Household Carbon Footprint
https://coolclimate.berkeley.edu/maps-2050

21.1 Residential Energy Use
https://www.eia.gov/energyexplained/use-of-energy/homes.php

Data – Additional Research Links

http://www.fao.org/economic/ess/ess-economic/gdpagriculture/en/

http://www.multpl.com/us-population-growth-rate

https://data.worldbank.org/indicator/NY.GDP.MKTP.KD?end=2017&start=1970

https://hbr.org/2018/06/25-years-ago-i-coined-the-phrase-triple-bottom-line-heres-why-im-giving-up-on-it

https://medium.com/@designforsustainability/redesigning-economics-based-on-ecology-35a82f751f25

https://resource.co/article/only-nine-percent-world-economy-circular-claims-global-circularity-report-12366

https://rhg.com/research/preliminary-us-emissions-estimates-for-2018/

https://www.axios.com/newsletters/axios-generate-ee0c7e82-559f-441b-b1b1-0755039dbf37.html

https://www.bloomberg.com/news/articles/2019-01-17/from-greenspan-to-yellen-economic-brain-trust-backs-carbon-tax

https://www.circularity-gap.world/report

https://www.climatelevels.org/

https://www.infoplease.com/world/population-statistics/total-population-world-decade-1950-2050

https://www.jacobinmag.com/2019/03/green-new-deal-financing-fossil-fuels

https://www.linkedin.com/pulse/circular-economy-save-world-whypart-1-2-kofi-a-adu-gyamfi/

https://www.newyorker.com/news/news-desk/the-false-choice-between-economic-growth-and-combatting-climate-change

https://www.nytimes.com/2019/01/08/climate/greenhouse-gas-emissions-increase.html

https://www.nytimes.com/2019/01/10/climate/ocean-warming-climate-change.html

https://www.nytimes.com/2019/03/15/opinion/sunday/elizabeth-warren-president-2020.html

https://www.nytimes.com/interactive/2018/12/07/climate/world-emissions-paris-goals-not-on-track.html

https://www.nytimes.com/interactive/2019/02/06/climate/fourth-hottest-year.html

https://www.thebalance.com/us-gdp-by-year-3305543

https://www.theguardian.com/commentisfree/2019/mar/15/capitalism-destroying-earth-human-right-climate-strike-children

https://www.theguardian.com/commentisfree/2019/mar/18/ending-climate-change-end-capitalism

https://www.theguardian.com/environment/2019/jan/08/carbon-emissions-trump-agenda-climate-change

https://thischangeseverything.org/book/

https://www.virgin.com/richard-branson/simple-idea-whip-climate-change

https://www.washingtonpost.com/news/opinions/wp/2019/01/02/feature/opinion-here-are-11-climate-change-policies-to-fight-for-in-2019/

https://www.washingtonpost.com/opinions/more-studies-show-terrible-news-for-the-climate-we-should-be-alarmed/2019/01/16/93da73d6-183f-11e9-9ebf-c5fed1b7a081_story.html

https://sites.psu.edu/math033fa17/2017/10/08/meatless-mondays-do-they-really-help/

Agriculture

https://phys.org/news/2018-09-climate-reshape-world-agricultural.html

https://medium.com/wedonthavetime/guest-blog-post-how-technology-in-agribusiness-affects-climate-change-a-concrete-solution-88505699174b

https://asi.ucdavis.edu/programs/ucsarep/about/what-is-sustainable-agriculture

https://www.bloomberg.com/news/articles/2019-05-01/beyond-meat-ipo-raises-241-million-as-veggie-foods-grow-fast

https://www.scientificamerican.com/article/lab-grown-meat/

Cost and Benefits of Apollo Program

https://christopherrcooper.com/blog/apollo-program-cost-return-investment/

https://www.investopedia.com/financial-edge/0812/the-roi-of-space-exploration.aspx

Education

Reimagining Education for a Sustainable Future
https://www.cusp.ac.uk/wp-content/uploads/09-Jonathan-Rowson-online.pdf

Historical References

Cost of WWII and ROI

The Cost of U.S. Wars Then and Now. Though it lasted fewer than four years, World War II was the most expensive war in United States history. Adjusted for inflation to today's dollars, the war cost over $4 trillion and in 1945, the war's last year, defense spending comprised about 40% of GDP.
https://m.warhistoryonline.com/history/cost-u-s-wars-now.html

America's involvement in World War II had a significant impact on the economy and workforce of the United States. American factories were retooled to produce goods to support the war effort and almost overnight the unemployment rate dropped to around 10%.
http://www.iptv.org/iowapathways/artifact/impact-world-war-ii-us-economy-and-workforce

As the Cold War unfolded in the decade and a half **after** World War II, the United States experienced phenomenal **economic** growth. The war brought the return of prosperity, and in the postwar period the United States consolidated its position as the world's richest country. ... The growth had different sources.
http://www.let.rug.nl/usa/outlines/history-1994/postwar-america/the-postwar-economy-1945-1960.php

Nuclear

https://www.npr.org/2019/05/08/720728055/this-company-says-the-future-of-nuclear-energy-is-smaller-cheaper-and-safer

https://www.forbes.com/sites/kensilverstein/2019/05/10/if-nuclear-energy-is-replaced-by-natural-gas-say-goodbye-to-climate-goals/#12398e920169

https://www.sckcen.be/en/SCKCEN_for_you/Energy/Gen_IV_re actors

https://www.sckcen.be/en/Technology_future/New_reactors

Nuclear Fusion

https://www.newsweek.com/nuclear-fusion-plasma-problem-solved-clean-energy-1117497

https://www.iaea.org/newscenter/news/fusion-energy-in-the-21st-century-status-and-the-way-forward

Oil Industry Expertise/Technology/Infrastructure for Permanent CO2 Sequestration

Exxon announced this week that it would invest US $100 million over ten years to research and develop advanced lower-emissions technologies with the U.S. Department of Energy's National Renewable Energy Laboratory and National Energy Technology Laboratory. The investment will support research and collaboration into ways to bring biofuels and **carbon capture and storage** to commercial scale across the transportation, power generation, and industrial sectors. $100 million over 10 years is a drop in the bucket, but it's a start. https://www.nasdaq.com/article/exxon-presents-its-very-own-solution-to-climate-change-cm1147019

Occidental wants to use the technology to find a sustainable supply of carbon dioxide that it can use to inject in its oil fields to increase pressure and extract more oil while also sequestering the carbon. The company is already the largest injector in the industry, but it now re-injects carbon that was found in natural underground deposits — providing little or no environmental benefit. By recycling carbon taken from the air, it hopes to bury as much carbon as its fuels emit, or even more. As an added benefit, there is a federal tax credit for sequestering carbon.
https://www.nytimes.com/2019/04/07/business/energy-environment/climate-change-carbon-engineering.html

While taking carbon directly from the air and sequestering it in rocks is far from a feasible scenario, capturing it at the source — power plants and refineries — is something that's already being tested at some commercial operations. There, carbon dioxide from burning or refining fuel is separated out and stored as a liquid for transport. Eventually, it will be injected into deep saline aquifers or spent oil wells and trapped, keeping it out of the atmosphere. Though around 30 million metric tons (Mt) of carbon dioxide annually are stored in this way today, scientists estimate that up to 1,000 Mt, or one billion tons, will need to be stored each year to counteract the effects of climate change. http://blogs.discovermagazine.com/d-brief/2018/09/04/carbon-capture-sequestration-storage-pipeline/#.XNWpGI5Kg2w

Theoretically, oil and gas giants could buy credits for sequestering the carbon equal to what their industries are releasing, making it a wash. Which would mean continuing the environmental destruction that comes with drilling and digging for resources we already have the capacity to replace with alternatives like solar and wind. It's delaying the inevitable. But the backing of these powerful industries could spur development of Negative Emissions Technologies (e.g. direct air capture) that could help solve the climate crisis.
https://www.wired.com/story/carbon-capture-is-messy-and-fraughtbut-might-be-essential/

Big innovations can happen when CEO bonuses are tied to emissions reductions. For a decade Statoil/Equinor has been perfecting methods of injecting carbon dioxide deep into the earth for permanent sequestration. It's now investing in solar farms in Brazil, and is putting to work its offshore engineering know-how with the world's first full-scale floating wind turbines at a project in the North Sea called Hywind. The size of the Eiffel Tower and producing 6 megawatts each, the turbines stand on buoyant bases that are inspired by decades of work building spar-type floating oil platforms. https://www.forbes.com/sites/christopherhelman/2018/10/12/norways-equinor-shows-big-oil-can-survive-putting-a-price-on-carbon/#68311a645c05

Isaac Asimov on Over Population

"I like to use what I call my bathroom metaphor. If two people live in an apartment, and there are two bathrooms, then they both have freedom of the bathroom. You can go to the bathroom anytime you want, stay as long as you want, for whatever you need. And everyone believes in freedom of the bathroom. It should be right there in the constitution. But if you have twenty people in the apartment and two bathrooms, then no matter how much every person believes in freedom of the bathroom, there's no such thing. You have to set up times for each person, you have to bang on the door, 'Aren't you through yet?' and so on."

And Asimov concluded with… "In the same way, democracy cannot survive overpopulation. Human dignity cannot survive overpopulation. Convenience and decency cannot survive overpopulation. As you put more and more people into the world, the value of life not only declines, it disappears. It doesn't matter if someone dies, the more people there are, the less one individual matters."

Personal – Becoming Crew

CO_2 Emissions from Typical Household Activities
https://www.ccfpd.org/Portals/0/Assets/PDF/Facts_Chart.pdf

This link provides information about companies who are selling 100% recycled products. https://greenbusinessbureau.com/blog/8-companies-that-only-sell-100-percent-recycled-products/ By purchasing from folks who are making products only from recycled content, we reduce the need to pillage more from Spaceship Earth's storeroom.

Here's a recommended site that helps people understand their personal power in tackling climate change; shows people the best responses to climate change; and inspires people to action to help them reduce their carbon footprint, adapt to climate change, and embrace a low carbon culture. https://www.52climateactions.com/actions-list

Neighborhood Climate Action Planning Handbook from Portland State University https://pdxscholar.library.pdx.edu/cgi/viewcontent.cgi?referer=&httpsredir=1&article=1042&context=usp_murp

Here's a recommended program: https://www.greenbiz.com/article/cool-through-community-climate-action-moonshot. A neighborhood Cool Block begins with an individual accepting the invitation to become a leader on the block where they live. Everything after that is an interconnected action-based exercise that takes place simultaneously in their household, with others on their block, and in partnership with their neighborhood or city.

Technologies

Additive Manufacturing http://additivemanufacturing.com/basics/

Autonomous Automobiles https://www.ornl.gov/ https://localmotors.com/

Big Data and Data Analytics https://www.ibm.com/it-infrastructure/solutions/big-data https://www.sas.com/en_us/whitepapers/tdwi-operationalizing-embedding-analytics-for-action-108112.html

Blockchain https://www.coindesk.com/information/what-is-blockchain-technology https://blockgeeks.com/guides/what-is-blockchain-technology/

Drawdown https://www.youtube.com/watch?v=RlowjpqY8QQ

Geoengineering https://monthlyreview.org/2018/09/01/making-war-on-the-planet/

Internet of Things https://digital.pwc.com/content/pwc-digital/en/products/connected-solutions.html https://www.lanner-america.com/blog/5-ways-internet-things-help-combat-climate-change/

Mass Transit https://www.bloomberg.com/news/articles/2019-05-15/in-shift-to-electric-bus-it-s-china-ahead-of-u-s-421-000-to-300

Satellite Imaging and Communications https://www.photonics.com/Articles/Hyperspectral_Imager_Will_Communicate_Climate/a64260 https://climate.nasa.gov/

Sensors https://www.fastcompany.com/90281427/to-save-agriculture-from-climate-change-we-need-better-weather-forecasting https://www.weforum.org/agenda/2018/08/how-technology-is-driving-new-environmental-solutions/ https://www.noaa.gov/education/resource-collections/climate-education-resources/climate-monitoring

Sharing Technologies http://www.chinadaily.com.cn/a/201811/10/WS5be62393a310eff303287c41.html https://bigthink.com/technology-innovation/climate-change-projects

Quotes

"Wasteful consumerism is a contemporary evil. We need to get back to economic, plain, simple and useful design." – Terence Conran, legendary British designer

"People with targets and jobs dependent upon meeting them will probably meet the targets – even if they have to destroy the enterprise to do it." – W. Edwards Deming

"Time is the scarcest resource and unless it is managed, nothing else can be managed." – Peter Drucker

"What can be counted doesn't always count, and not everything that counts can be counted." – Albert Einstein

"Everything is energy and that's all there is to it. This is not philosophy. This is physics." – Albert Einstein

"Our common home is being pillaged, laid waste and harmed with impunity. Cowardice in defending it is a grave sin." – Pope Francis

"If the success or failure of this planet and of human beings depended on how I am and what I do... How would I be? What would I do?" – R. Buckminster Fuller

"We are called to be architects of the future, not its victims." – R. Buckminster Fuller

"We are not going to be able to operate our Spaceship Earth successfully, nor for much longer, unless we see it as a whole spaceship and our fate as common. It has to be everybody or nobody." – R. Buckminster Fuller

"Tell me how you measure me, and I will tell you how I will behave." – Eliyahu M. Goldratt

"If you don't collect any metrics, you're flying blind. If you collect and focus on too many, they may be obstructing your field of view." – Scott M. Graffius, Agile Scrum

"Today, nearly everyone can point to a way that they are personally witnessing and are being affected by the impacts of a changing climate in the places where they live." – Katharine Hayhoe

"We have entered the Golden Age of Design." – David Houle 2016 AICAD keynote

"We all live down-wind, down-stream, and down-time from everybody else. We need to think, design, and operate in a systems model – a holistic context of dynamic equilibrium." – Houle & Rumage, This Spaceship Earth

"Entering the 21st century, the vast majority of humanity has been living in a world with a footprint greater than the planet's capacity. Clearly, humanity cannot continue to live beyond the Earth's means... we must alter the trajectory, the thinking, and the practices that have placed us in this mission critical state." – Houle and Rumage, This Spaceship Earth

"We are in a changing, dynamic, planetary situation that has never existed during modern humanity's time on Earth. We are convinced that this calls for a new consciousness, a planetary crew consciousness." – Houle and Rumage, This Spaceship Earth

"I've never been at a climate conference where people say, 'that happened slower than I thought it would.'" – Christina Hulbe, Climatologist

"For peace to reign on earth, humans must evolve into new beings who have learned to see the whole first." – Immanuel Kant

"Our most basic common link is that we all inhabit this small planet, we all breathe the same air, we all cherish our children's futures, and we are all mortal." – John F. Kennedy

"People are unlikely to jettison an unworkable paradigm, despite many indications it is not functioning properly, until a better paradigm can be presented." – Thomas Kuhn

"We shouldn't give up hope until we've done all we can." – Bob Leonard

"There's a mechanistic worldview that unreservedly exploits the natural world because it attributes no inherent value to nature. This implies that the ecological crisis is nothing more than a technological problem, or an economic problem, or a political problem. It is also a collective spiritual crisis, and a potential turning point in our history." – David Loy, Eco-dharma: Buddhist Teachings for the Ecological Crisis

"Environment and economy are two sides of the same coin. If we cannot sustain the environment, we cannot sustain ourselves." – Wangari Maathai

"There is nothing so difficult to plan, more uncertain of success, nor dangerous to manage than the creation of a new order of things. For the initiator has the enmity of all those who would profit by the preservation of the old institutions and merely lukewarm defenders in those who would gain by the new ones." – Niccolò Machiavelli

"We stand at a new threshold, and the key questions of our day are those posed by the intersection of design and life itself." – Bruce Mau

"We are destroying the natural world. And that means that we are destroying ourselves." – Chris Martenson, Peak Prosperity

"There are no passengers on Spaceship Earth. We are all crew." – Marshall McLuhan

"What mankind must know is that human beings cannot live without Mother Earth, but the planet can live without humans. We won't have a society if we destroy the environment." – Margaret Mead

"Only in economics is endless expansion seen as a virtue. In biology, it is called cancer." – David Pilling

"If it can't be reduced, reused, repaired, rebuilt, refurbished, refinished, resold, recycled or composted, then it should be restricted, redesigned or removed from production." – Pete Seeger

"Nature shrinks as capital grows. The growth of the market cannot solve the very crisis it creates." – Vandana Shiva, Soil Not Oil

"The secret of change is to focus all of your energy, not on fighting the old, but on building the new." – Socrates

"… if we want to know the secrets of the universe, we should focus on the non- physical aspects rather than physical ones, that will speed up the inventions." – Nikola Tesla

"Adults keep saying, 'We owe it to the young people to give them hope.' I don't want your hope. I don't want you to be hopeful. I want you to panic. I want you to feel the fear I feel every day. And then I want you to act. I want you to act as you would in a crisis. I want you to act as if our house is on fire. Because it is." – Greta Thunberg, 16-year-old climate activist

"Any serious restructuring of business must directly attack the organization and the entire system of power based on it." – Alvin Toffler

"The illiterate of the 21st Century are not those who cannot read and write but those who cannot learn, unlearn and relearn." – Alvin Toffler

"We should use Nature as the measure, using Nature's wisdom as a template for our economic systems." – Douglas Tompkins

"I've just become less sanguine about how we were going to fix the world by tomorrow. That's clearly not going to happen, because too many people have too much of a stake in what's wrong with the world, and I believe it's all fear. I don't believe anybody, given the full choice – except for sociopaths – would prefer to be operating on a basis of greed and acquisition, because everybody knows that the actual possession of things themselves does not generate any long-term satisfaction. Everybody knows this, and yet people keep chasing the carrot, even though they sort of know that it's tied to their own heads and they'll never get it, they still keep chasing it because they don't know how not to." – Peter Tork, member of The Monkees

"The journey of a thousand miles begins with one step." – Lao Tzu

"Gentlemen, the President just told us that we need to do something that has never been done before, using technology that has yet to be invented, to do it in a short amount of time and that safety is a top priority. Let's get to work!" – attributed to James Webb, leader of NASA from 1961-1968, shortly after President Kennedy's 1961 speech about landing a man on the moon by the end of the decade.

"The major advances in civilization are processes that all but wreck the societies in which they occur." – Alfred North Whitehead

"There is a tendency to think that what we see in the present moment will continue. We forget how often we have been astonished by the sudden crumbling of institutions, by extraordinary changes in people's thoughts, by unexpected eruptions of rebellion against tyrannies, by the quick collapse of systems of power that seemed invincible." – Howard Zinn

Footnotes

1.1   https://www.ipcc.ch/sr15/

2.1   https://www.climate.gov/news-features/understanding-climate/climate-change-atmospheric-carbon-dioxide

2.2   https://pubs.giss.nasa.gov/abs/ha02700w.html

3.1   http://www.ces.fau.edu/nasa/module-2/how-greenhouse-effect-works.php

3.2   https://recoverusa.com/industrial-waste-management/

3.3   http://www.worldbank.org/en/topic/urbandevelopment/brief/solid-waste-management

3.4   https://www.climate.gov/news-features/understanding-climate/climate-change-atmospheric-carbon-dioxide

3.5   https://www.eia.gov/electricity/sales_revenue_price/pdf/table5_a.pdf

4.1   https://www.circularity-gap.world/

7.1   https://www.sciencedirect.com/science/article/abs/pii/S0305750X16304867

7.2   https://soulofmoney.org/

8.1   https://space.nss.org/settlement/nasa/spaceresvol4/newspace3.html

8.2   https://www.moneycrashers.com/leading-lagging-economic-indicators/

8.3   www.biologicaldiversity.org/programs/biodiversity/elements_of_biodiversity/extinction_crisis/

9.1   https://www.focus-economics.com/blog/the-largest-economies-in-the-world

9.2   https://www.internetworldstats.com/stats8.htm

9.3   https://www.cdc.gov/media/releases/2018/p1108-cigarette-smoking-adults.html

9.4   https://www.clcouncil.org/our-plan/

9.5   https://www.thoughtco.com/top-new-deal-programs-104687

10.1   https://www.schroders.com/en/sysglobalassets/digital/us/pdfs/the-impact-of-climate-change.pdf

10.2   https://www.weforum.org/agenda/2019/01/these-are-the-biggest-risks-facing-our-world-in-2019/

10.3   https://www.technologyreview.com/s/613328/

10.4   https://www.who.int/air-pollution/news-and-events/how-air-pollution-is-destroying-our-health

10.5   https://www.technologyreview.com/s/613328/the-one-number-you-need-to-know-about-climate-change/

10.6   https://www.overshootday.org/

10.7   https://data.worldbank.org/

10.8   https://www.epa.gov/sites/production/files/2016-05/global_emissions_sector_2015.png

11.1   https://data.worldbank.org/

11.2   https://www.ucsusa.org/clean-energy/coal-and-other-fossil-fuels/environmental-impacts-of-natural-gas

11.3   https://www.nrdc.org/experts/danielle-droitsch/time-us-end-fossil-fuel-subsidies

11.4   https://www.rollingstone.com/politics/politics-news/fossil-fuel-subsidies-pentagon-spending-imf-report-833035/

11.5 https://www.imf.org/en/Publications/WP/Issues/2019/05/02/Global-Fossil-Fuel-Subsidies-Remain-Large-An-Update-Based-on-Country-Level-Estimates-46509

11.6 https://www.popularmechanics.com/science/energy/a21756137/renewables-50-percent-energy-2050/

11.7 https://www.c2es.org/content/renewable-energy/

11.8 https://www.weforum.org/agenda/2018/01/renewables-will-be-equal-or-cheaper-than-fossil-fuels-by-2020-according-to-research

11.9 https://greenliving.lovetoknow.com/Efficiency_of_Wind_Energy

11.10 https://news.energysage.com/solar-farms-start-one/

11.11 http://www.windustry.org/how_much_do_wind_turbines_cost

11.12 http://www.ren21.net/wp-content/uploads/2018/06/GSR_2018_Highlights_final.pdf

11.13 https://askjaenergy.com/iceland-renewable-energy-sources/geothermal-different-utilization/

11.14 https://www.eniday.com/en/sparks_en/geothermal-energy-iceland/

11.15 https://www.greenfacts.org/en/geothermal-energy/l-2/index.htm#0

11.16 https://www.statista.com/statistics/476267/global-capacity-of-marine-energy/

11.17 https://www.climaterealityproject.org/blog/why-marine-energy-wave-future

11.18 https://cen.acs.org/articles/91/web/2013/04/Nuclear-Power-Prevents-Deaths-Causes.html

11.19   http://time.com/5255663/chernobyl-disaster-book-anniversary/

11.20   https://www.who.int/ionizing_radiation/a_e/fukushima/faqs-fukushima/en/

11.21   https://www.ucsusa.org/clean-energy/coal-and-other-fossil-fuels/hidden-cost-of-fossils

11.22   https://www.ourenergypolicy.org/wp-content/uploads/2019/03/Become-a-Nuclear-Safety-Expert-Rev.pdf

11.23   https://www.mofa.go.jp/region/n-america/us/security/fact0604.pdf

11.24   https://www.bloomberg.com/graphics/2019-nuclear-waste-storage-france/

11.25   https://www.bloomberg.com/graphics/2019-nuclear-waste-storage-france/

11.26   http://www.world-nuclear.org/information-library/nuclear-fuel-cycle/nuclear-power-reactors/advanced-nuclear-power-reactors.aspx

11.27   http://www.world-nuclear.org/information-library/nuclear-fuel-cycle/nuclear-power-reactors/generation-iv-nuclear-reactors.aspx

11.28   https://www.forbes.com/sites/jamesconca/2016/03/24/is-nuclear-power-a-renewable-or-a-sustainable-energy-source/#5ac45559656e

11.29   http://eas.caltech.edu/engenious/twelve/idea_flow

11.30   https://earthsky.org/earth/space-based-solar-energy-power-getting-closer-to-reality

11.31   https://www.nbcnews.com/mach/science/solar-farms-space-could-be-renewable-energy-s-next-frontier-ncna967451

11.32   https://www.smithsonianmag.com/innovation/whats-next-solar-energy-how-about-space-180961008/#fwPtMhjApfS4X2Jp.99

11.33   https://www.energy.ca.gov/tour/solarstar/

11.34   https://www.smithsonianmag.com/innovation/whats-next-solar-energy-how-about-space-180961008

11.35   https://www.axios.com/the-key-to-unlocking-wind-and-solar-making-it-last-74ce3a20-ab7e-4410-a883-011abac6eafc.html

11.36   https://electrek.co/2018/05/11/tesla-giant-battery-australia-reduced-grid-service-cost/

11.37   https://www.iter.org/

11.38   https://www.nytimes.com/2018/06/26/opinion/farming-organic-nature-movement.html

11.39   https://www.washingtonpost.com/national/health-science/burgers-grown-in-a-lab-are-heading-to-your-plate-will-you-bite/2018/09/07/1d048720-b060-11e8-a20b-5f4f84429666_story.html

11.40   https://www.fastcompany.com/40579068/there-were-3-1-million-electric-vehicles-on-the-road-in-2017

11.41   https://cleantechnica.com/2019/05/30/iea-predicts-250-million-evs-on-the-road-by-2030/

11.42   https://www.sciencedirect.com/science/article/pii/S0001457518300873

11.43   http://fortune.com/2016/03/13/cars-parked-95-percent-of-time/

11.44   https://www.weforum.org/agenda/2018/05/how-technology-is-driving-a-fourth-wave-of-environmentalism/

11.45   https://www.theguardian.com/world/2019/may/23/china-factories-releasing-thousands-of-tonnes-of-illegal-cfc-gases-study-finds

11.46   https://www.climatechangenews.com/2019/05/27/eu-plans-satellite-fleet-monitor-co2-every-country/

11.47   https://science.howstuffworks.com/environmental/green-science/clean-coal.htm

11.48   https://www.nytimes.com/2019/04/07/business/energy-environment/climate-change-carbon-engineering.html

11.49
https://projects.ncsu.edu/project/treesofstrength/treefact.htm

11.50
www.researchgate.net/publication/329269125_Artificial_trees_for_atmospheric_carbon_capture

11.51   https://www.ucsusa.org/global-warming/science-and-impacts/science/each-countrys-share-of-co2.html

11.52   https://www.wri.org/blog/2018/09/6-ways-remove-carbon-pollution-sky

11.53   https://www.cell.com/joule/fulltext/S2542-4351(18)30225-3

11.54   https://www.eia.gov/outlooks/aeo/data/browser/#/

11.55   https://www.wri.org/blog/2018/09/6-ways-remove-carbon-pollution-sky

11.56   http://www.geoengineeringmonitor.org/2018/05/enhanced-weathering-factsheet/

12.1   https://sloanreview.mit.edu/article/transforming-manufacturing-and-supply-chains-with-4d-printing/

12.2   https://www.interaction-design.org/literature/article/holistic-design-design-that-goes-beyond-the-problem

12.3   http://www.mcdonoughpartners.com/

12.4
https://extension.tennessee.edu/publications/Documents/W235-F.pdf

15.1   https://www.climate.gov/news-features/understanding-climate/climate-change-global-sea-level

15.2 https://www.nationalgeographic.com/environment/2019/01/greeland-ice-melting-four-times-faster-than-thought-raising-sea-level/

15.3   http://worldpopulationreview.com/us-cities/miami-population/

15.4   https://www.thestreet.com/personal-finance/real-estate/us-cities-most-in-danger-of-rising-seas-14839105
https://www.zillow.com/research/ocean-at-the-door-21931/

15.5 https://static01.nyt.com/packages/pdf/tbooks/Climate_Refugees_NYTimes_050317.pdf

15.6 https://www.un.org/sustainabledevelopment/blog/2019/06/lets-talk-about-climate-migrants-not-climate-refugees/

16.1   https://www.worldpopulationbalance.org/faq

16.2   https://news.un.org/en/story/2019/01/1029592

16.3 https://www.theguardian.com/environment/2017/jul/12/want-to-fight-climate-change-have-fewer-children

16.4   https://populationmatters.org/the-facts/climate-change

16.5 https://www.biologicaldiversity.org/programs/population_and_sustainability/pdfs/OSUCarbonStudy.pdf

16.6   https://www.drawdown.org/

16.7 https://www.cdc.gov/mmwr/preview/mmwrhtml/mm4847a1.htm

16.8   https://www.fastcompany.com/90315700/meet-the-women-deciding-not-to-have-kids-because-of-climate-change

16.9 https://www.biologicaldiversity.org/programs/population_and_sustainability/climate/index.html

19.1   https://www.nytimes.com/2019/07/07/business/energy-environment/florida-solar-power.html

19.2   https://greywateraction.org/greywater-codes-and-policy/

20.1   https://arstechnica.com/science/2015/05/megacities-demand-lots-of-energy-but-result-in-lots-of-gdp/

20.2   http://go.euromonitor.com/rs/805-KOK-719/images/MegacitiesExtract.pdf

20.3   https://www.weforum.org/agenda/2018/10/these-are-the-megacities-of-the-future/

20.4 https://www.salon.com/2014/01/08/this_map_shows_how_suburban_sprawl_is_destroying_the_environment/

20.5   https://www.farandwide.com/s/public-transit-systems-ranked-c5d839d8a48d4da3

20.6   https://www.babcockranch.com/about-us/

21.1   https://www.greenbiz.com/article/cool-through-community-climate-action-moonshot

21.2   https://www.visualcapitalist.com/what-uses-the-most-energy-home/

21.3   https://www.commonobjective.co/article/can-fashion-stop-climate-change

21.4   https://www.agric.wa.gov.au/climate-change/composting-avoid-methane-production

21.5   https://www.cusp.ac.uk/wp-content/uploads/09-Jonathan-Rowson-online.pdf

23.1
https://www.nytimes.com/2005/01/30/books/review/collapse-how-the-world-ends.html

23.2
https://www.worldbank.org/en/topic/climatechange/overview

Biographies

David Houle

David Houle is a futurist, strategist and speaker. Houle spent more than 20 years in media and entertainment. He worked at NBC, CBS, and was part of the senior executive team that created and launched MTV, Nickelodeon, VH1 and CNN Headline News.

Houle has won a number of awards. He won two Emmys, the prestigious George Foster Peabody award, and the Heartland award for "Hank Aaron: Chasing the Dream." He was also nominated for an Academy Award. He is the Futurist in Residence at the Ringling College of Art + Design, the Co-founder of The Sarasota Institute – a 21st century think tank, and the Honorary Futurist at the Future Business School of China.

He has delivered 1000+ speeches, presentations and corporate retreats on six continents and sixteen countries. He is often called "the CEOs' Futurist" having spoken to or advised 4,500+ CEOs and business owners in the past eleven years. He was invited to present "This Spaceship Earth" to scientists at the NASA Goddard Space Flight Center and at the EPA.

Houle coined the phrase the Shift Age and has written extensively about the future and the future of energy. This is his eighth book. His primary web site is https://davidhoule.com/

Bob Leonard

Bob is an author researcher, editor and ghostwriter. He works with David as Editor and Researcher. He's also the Chief Content Officer at This Spaceship Earth, a climate change nonprofit founded by David Houle and Tim Rumage.

Following a career in B2B technology marketing at Interleaf, GTE and EMC, Bob decided to follow his passion and focus his attentions on the existential crisis that is climate change. Hired as the Director of Operations at This Spaceship Earth, Bob was subsequently promoted to Chief Content Officer.